Measurement Science for Modern Physics

Other related titles:

You may also like

- PBTE099 | Loh | Metrology for 5G and Emerging Wireless Technologies | 2021
- PBHT018 | Tunbridge | Lord Kelvin: His influence on electrical measurements and units | 1992
- PBEL007 | Fantom | Radio Frequency and Microwave Power Measurement | 1990

We also publish a wide range of books on the following topics:
Applied Physics
Agricultural Engineering & Technology
Built Environment
Computing and Networks
Control, Robotics and Sensors
Electrical Regulations
Electromagnetics and Radar
Energy Engineering
Healthcare Technologies
History and Management of Technology
IET Codes and Guidance
Materials, Circuits and Devices
Model Forms
Nanomaterials and Nanotechnologies
Production, Design and Manufacturing
Security
Telecommunications
Transportation

All books are available in print via https://shop.theiet.org or as eBooks via our Digital Library https://digital-library.theiet.org.

IET APPLIED PHYSICS SERIES 11

Measurement Science for Modern Physics

Masatoshi Kajita

The Institution of Engineering and Technology

About the IET

This book is published by the Institution of Engineering and Technology (The IET).

We inspire, inform, and influence the global engineering community to engineer a better world. As a diverse home across engineering and technology, we share knowledge that helps make better sense of the world, to accelerate innovation and solve the global challenges that matter.

The IET is a not-for-profit organization. The surplus we make from our books is used to support activities and products for the engineering community and promote the positive role of science, engineering, and technology in the world. This includes education resources and outreach, scholarships and awards, events and courses, publications, professional development and mentoring, and advocacy to governments.

To discover more about the IET please visit https://www.theiet.org/.

About IET books

The IET publishes books across many engineering and technology disciplines. Our authors and editors offer fresh perspectives from universities and industry. Within our subject areas, we have several book series steered by editorial boards made up of leading subject experts.

We peer review each book at the proposal stage to ensure the quality and relevance of our publications.

Get involved

If you are interested in becoming an author, editor, series advisor, or peer reviewer, please visit https://www.theiet.org/publishing/publishing-with-iet-books/ or contact author_support@theiet.org.

Discovering our electronic content

All of our books are available online via the IET's Digital Library. Our Digital Library is the home of technical documents, eBooks, conference publications, real-life case studies, and journal articles. To find out more, please visit https://digital-library.theiet.org.

In collaboration with the United Nations and the International Publishers Association, the IET is a Signatory member of the SDG Publishers Compact. The Compact aims to accelerate progress to achieve the Sustainable Development Goals (SDGs) by 2030. Signatories aspire to develop sustainable practices and act as champions of the SDGs during the Decade of Action (2020–30), publishing books and journals that will help inform, develop, and inspire action in that direction.

In line with our sustainable goals, our UK printing partner has FSC accreditation, which is reducing our environmental impact on the planet. We use a print-on-demand model to further reduce our carbon footprint.

First published: 2026

The Institution of Engineering and Technology
Futures Place,
Kings Way,
Stevenage,
Herts, SG1 2UA,
United Kingdom

www.theiet.org

British Library Cataloguing in Publication Data
A catalogue record for this product is available from the British Library

ISBN 978-1-80705-168-6 (hardback)
ISBN 978-1-80705-169-3 (PDF)
ISBN 978-1-80705-170-9 (ePub3)

Typeset in India by MPS Limited

Cover Image: Erik Simonsen/Photodisc via Getty Images

I learned the fundamentals of laser spectroscopy from Prof. T. Shimizu as a graduate course student at the University of Tokyo. After joining the Communications Research Laboratory (since 2004, renamed National Institute of Information and Communications Technology (NICT)), I learned the fundamentals of precision measurement of atomic transition frequencies from Prof. K. Nakagiri, Prof. S. Urabe, Dr. T. Morikawa, and Dr. M. Hosokawa. I recognized the critical role of precision frequency measurement for all fields of physics after joining "Fundamental Physics Using Atoms," organized by Prof. N. Sasao, Prof. K. Asahi, Prof. Y. Sakemi, and Prof. K. Sugiyama. Prof. T. Aoki and Prof. Y. Torii gave me a chance to continue research at the University of Tokyo after retirement from NICT. Prof. S. Goka gave me a chance for research activity in Tokyo Metropolitan Univ. after Apl. 2026.

I appreciate all the people listed above for their help in the publication of this book.

Contents

Preface

The purpose of physical research is the establishment of physical laws based on observed phenomena. Theorists predict the phenomena in the future by calculation based on the established laws. Experimentalists confirm the utility of established laws by observing the phenomena under different conditions. Establishments of new physical laws are required with the discovery of new phenomena, which cannot be described by previous laws.

Note that the physical laws are generally given by the simplest descriptions, with which the observed phenomena are described without discrepancies. Some discrepancies with previous physical laws have been discovered with the (1) expansion of the observation area and (2) reduction of measurement uncertainties. The ancient people lived in a small area, and the physical phenomena were observed under limited conditions. Whilst they were living within an area of 50 km, they believed the Earth to be flat. Moving the distance of thousands of km from north to south, the height of constellations changed, and the spherical form of the Earth was recognized. Until the Middle Ages, they believed that the Sun and planets revolved around the Earth. There was no discrepancy with this idea, whilst there was uncertainty of time by 30 min. With the improvement of the accuracy of clocks, the observed positions of planets at certain times were clarified to be different from theoretical predictions. These discrepancies were solved with the idea that the Earth and other planets revolve around the Sun (Copernican theory). Galilei made a great contribution to the establishment of the Copernican theory by the drastic improvement of the accuracy of clocks from the discovery of the periodicity of pendulum swings. He also observed the Moon, Jupiter, and Sun using a handmade telescope, which was the first observation of phenomena outside the Earth. After the publication of Kepler's law, Newtonian mechanics was established.

To observe the phenomena of other planets using an accurate clock, the propagation time of light is not negligible. The speed of light was clarified to be finite from the observation of the periodical fluctuation of the revolution period of Io (one of the satellites of Jupiter). The speed of light in a vacuum was clarified to be constant for any observer with different motion velocities, which indicated the limit of the utility of Newtonian mechanics. The theory of relativity is a mechanics based on the constant speed of light, which indicates that Newtonian mechanics is approximately valid with the motion velocity much lower than the speed of light. Light has dual properties; electromagnetic wave and particle (called a photon). Each photon has energy proportional to its frequency. The wave natures of particles were clarified, which indicated that Newtonian mechanics is not valid to describe the phenomena of micro size (same order as the wavelength of matter waves). Atoms and molecules have discrete internal energy, and the state can be changed by absorbing or emitting electromagnetic waves (light, microwave) with discrete transition frequencies. An atomic clock is a clock based on the transition frequencies of atoms, which drastically reduces the measurement uncertainty of time and frequency. The relativistic effects with the progress of time were confirmed using atomic clocks. The development of lasers as light sources with ultra-narrow frequency bandwidth and uniform phase gave a revolution in experimental physics.

Until the nineteenth century, astronomical research was performed only for small fractions of stars in the Milky Way. After the observation of galaxies outside the Milky Way, the universe was clarified to be expanding since the massive explosion from a single point. The existence of dark energy and dark matter, whose identities are mysteries of modern physics, was clarified.

The accelerators of charged particles were constructed since the beginning of the twentieth century, with which particles are accelerated to speed close to that of light. A high-energy collision is required to observe the phenomena in nuclei because nuclear size is five orders smaller than atomic size (short wavelength of constructing particles) and the energy gap between different states is large. Protons were believed to be elemental particles until 1969, but it was clarified to be the binding of three quarks from the scattering of electrons that collided with ultra-high energy. Now the temperature of 4×10^{12} K was attained (plasma state of quarks and gluons), which is expected to be the reproduction of the state soon (order 10^{-10} s) after the birth of the universe. New particles have been discovered via pair production in high-energy collisions of particles.

The temperature below 100 K had never been attained until the end of the nineteenth century (liquid nitrogen was produced in 1883). The liquid helium with 4 K (−269 °C) was produced in 1908, and new phenomena with temperatures below 10 K were observed (superfluidity, superconductivity). After the development of lasers, laser cooling technology was developed to decelerate atoms to a kinetic energy below 10^{-3} K. The wavelength of atomic waves becomes longer with ultra-low kinetic energy, and several quantum effects were observed.

This book introduces the previous procedure of physics research based on the expansion of the observation area and the search for slight effects through precision measurement.

Acknowledgements

The research activity of the author has been supported by a Grant-in-Aid for Scientific Research (B) (Grant Nos. JP 17H02881 and JP20H01920) and a Grant-in-Aid for Scientific Research (C) (Grant Nos. JP 17K06483 and 16K05500) from the Japan Society for the Promotion of Science (JSPS). The author is highly appreciative of discussions with Y. Yano, T. Ido, N. Ohtsubo, A. Shinjo-Kihara, H. Hachisu, S. Nagano, M. Kumagai, S. Hayashi, N. Sekine, and M. Hara, all from the NICT, Japan, as well as K. Okada (Sophia University) and N. Kimura (National Institute for Fusion Science). T. Aoki and Y. Torii (University of Tokyo) gave me a chance to continue the research activity at the University of Tokyo after retirement from NICT. The author is grateful to J. Navas for giving the opportunity to write this book. Prof. S. Goka gave me a chance for research activity in Tokyo Metropolitan Univ. since April 2026.

About the author

Masatoshi Kajita is a guest professor at Tokyo Metropolitan University, Japan. He was born and raised in Nagoya, Japan. He earned his BS in applied physics from the University of Tokyo in 1981 and his PhD in physics in 1986. After working at the Institute for Molecular Science, he joined the Communications Research Laboratory in 1989 (later NICT). He was a guest professor at Université de Provence in 2009 and, after retiring from NICT in 2023, he became a guest researcher at the University of Tokyo. In April 2026, he took up his current position at Tokyo Metropolitan University.

Chapter 1

Development of physics research up to the nineteenth century

Abstract

New scientific results are obtained by the expansion of the measurement area and the reduction of measurement uncertainty. This chapter introduces the procedure of physics research since ancient times. In ancient times, the Earth was believed to be a flat. The spherical figure of the Earth was recognized by the dependence of the heights of the Sun or constellations on the latitude since the motion between Greece and the Middle East became active. The revolution of the Earth around the Sun was recognized after the accuracy of clocks was improved and the correlation between the observation time and the positions of planets was analyzed in detail. The speed of light was clarified to be finite by observing the phenomena on other planets using accurate clocks. In the nineteenth century, the ideas of atoms and molecules were established. Until the end of the nineteenth century, atoms were believed to be unit particles. But since the discovery of the electrons, it was clarified that atoms have a structure with a nucleus and electrons, which cannot be explained by previous physics. The establishment of quantum mechanics was required to describe the phenomena at microsize.

1.1 Introduction

The foundation of scientific research is the interpretation of the physical laws based on the measurement results. The reliability of the established laws has been confirmed by observing phenomena under other conditions. The established laws based on the measurements in some limited areas are not always universal. Note also that there are finite uncertainties with measurement results.

From the measurement results of $(x, y) = (0.3, 0.3 \pm 0.03)$, $(0.5, 0.5 \pm 0.05)$, $(0.7, 0.7 \pm 0.07), (1.0, 1.0 \pm 0.1)$, a simple interpretation of $y = x$ is given. However, we do not know whether this proportional relation is strictly valid or if it is an approximation that is valid with $[x] < 1$. While measuring within the region of $0 < x < 1$, it is difficult to distinguish the difference between $y = x$ and $y = \sin(x)$ as shown in Figure 1.1.

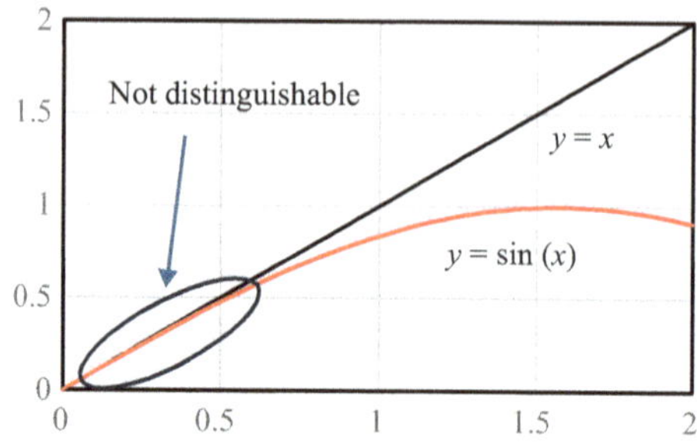

Figure 1.1 *Comparison of the functions* $y = x$ *and* $y = \sin(x)$. *Both functions are not distinguishable in* $0 < x < 0.5$, *but the difference is significant at larger* x.

Expanding the measurement region to $0 < x < 2$, the difference between the functions becomes significant, also with the measurement uncertainty of 10%. We can distinguish the difference between the functions also by reducing the measurement uncertainty down to 1% with the measurement results with $0.4 < x < 1$.

1.2 Early conceptions of the Earth's form

The development of physics is given by the expansion of the observation area. Observing the phenomena in a small area, we cannot know the truth. For example, the Earth was believed to be a flat until the sixth century BC. The life of ancient people was quite simple: just getting up at sunrise and going to sleep at sunset. The periodic motion of the Sun, Moon, and stars from east to west was recognized, although they did not know the cause. They did not recognize the dependence of the height of the Sun and stars on the latitude while they lived in an area smaller than 50 km.

Since the sixth century BC, the motion between Greece and the Middle East became active. Moving south by 1000 km, the constellations at the southern horizon were seen at higher positions, and the Earth was recognized as a sphere. This is the first example of how to know the real nature by expanding the observation area. The spherical shape of the Earth is also confirmed from the circular shadow of the Earth at a lunar eclipse. In the fifth century BC, nobody doubted the spherical figure of the Earth.

In 230 BC, Eratosthenes (Greek philosopher) estimated the radius of the Earth from the difference in the height of the Sun at the noon of the summer solstice in Alexandria and Syene (modern Aswan, 925 km south of Alexandria), as shown in Figure 1.2 [1]. There is a difference in the angles of shadows of vertical rods cast by the Sun ($\theta_{1,2}$) by $\theta_1 - \theta_2 = 0.04\pi$ radian. The circumference and radius of the Earth were estimated to be 46,250 and 7400 km, respectively. These values are 15% larger than the present values. The spherical figure of the Earth could not be recognized without the measurements of the height of the Sun or stars with low measurement uncertainty at two or more distant places.

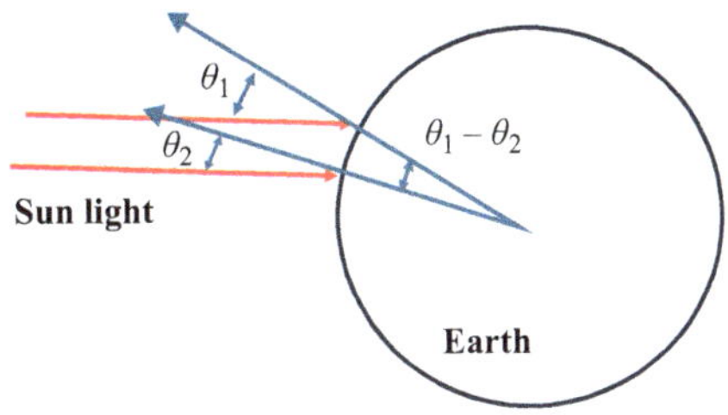

Figure 1.2 *Estimation of the radius of the Earth from the heights of the Sun at two places*

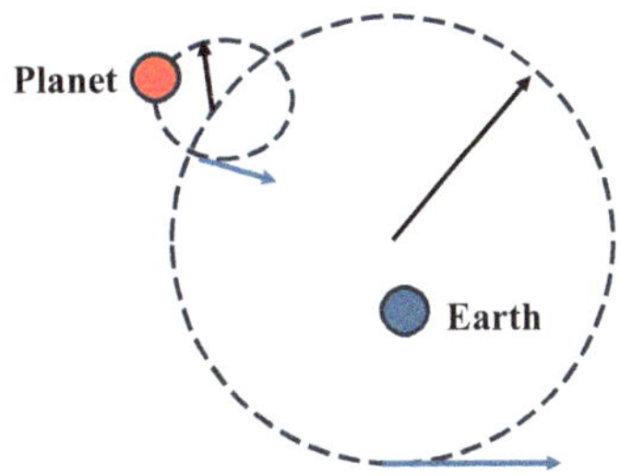

Figure 1.3 *The concept of Ptolemaic theory with double-circle revolution around the position shifted from the Earth*

1.3 The Ptolemaic theory of planetary motion

After the establishment of the spherical figure of the Earth, the motion of the Sun, Moon, planets, and stars from east to west was believed to indicate the revolution motion around the Earth. Their motions should be analyzed from the correlation between their positions and time. Detailed analysis was not possible when time was measured with sundials or water clocks, with which there was an uncertainty of 1 h per day.

However, motions of five planets (Mercury, Venus, Mars, Jupiter, and Saturn) were distinguished from stars because (i) they move on the same orbit as the Sun and (ii) their motion velocities are not constant. The occasional reverse motion of the planets was a serious mystery. In the second century, Ptolemy gave a hypothesis to describe the motions of planets (the Sun also was believed to be one of the planets) with the following two points (see Figure 1.3) [2].

1. Planets revolve around a point, which revolves around the Earth (then the occasional reverse motion is explained).
2. Planets revolve around a position that is shifted from the Earth (therefore, the motion velocities of the planets are not constant).

Although this theory is only the empirical one, the motion of planets was described without any discrepancies with the measurement uncertainty at that era.

The uncertainties of observation time were reduced after mechanical clocks were developed in the Middle Ages in Europe. Some revisions were given to the Ptolemaic theory to solve the discrepancies with observations, and the theory became a more complicated one.

1.4 Recognition of Earth's revolution around the Sun

The idea that the Earth revolves around the Sun was given in the third century BC by Aristarchus, but this idea had been forgotten until Copernicus published a new analysis in 1510. Considering that the Earth and planets are revolving around the Sun with different orbits, the motion velocity of planets observed on Earth depends on the relative positions between the Earth and the planets. The occasional reverse motion of planets is also possible with this simple hypothesis. It was called the Copernican theory. However, Copernican theory was not accepted because it did not give a significant improvement in the prediction of the positions of planets in comparison with Ptolemaic theory. The main problem with this theory was the assumption of the circle orbits of planets. The accuracy of observation of planets and stars was not high enough to get detailed information about the planets having elliptical orbits.

Galilei provided a step for the establishment of the Copernican theory. In 1581, Galilei discovered the periodicity of the pendulum's swing, which made a rapid improvement in the accuracy of clocks, and the motion of planets was observed with high accuracy.

Galilei made the first observation of the phenomena outside of the Earth using a handmade telescope. In 1609, he observed four satellites of Jupiter. He also observed the periodical change in the size of the observed Jupiter, which was explained by the periodic change in the distance to Jupiter due to the revolution of Jupiter and Earth in different orbits around the Sun. In 1612, he observed the rotation of the Sun, which inspired him that the Earth also rotates. It became easy to imagine the motion of the Earth via astronomical observations by Galilei.

The Copernican theory was finally established between 1609 and 1619 when Kepler published the following three laws of planetary motion [3].

1. The orbit of a planet is an ellipse with the Sun at one of the two foci.
2. A line segment joining a planet and the Sun sweeps out equal areas during equal time intervals.
3. The square of the orbital period of a planet is proportional to the cube of the semi-major axis of orbit.

Kepler's law was explained by Newtonian law afterward (Chapter 3.2). Ptolemaic theory is now denied, mainly because the double-circle revolution proposed by Ptolemaic theory was clarified to be impossible. The chance for giving the idea of elliptical orbit was given by Ptolemy's idea of revolution around the position shifted from the Earth.

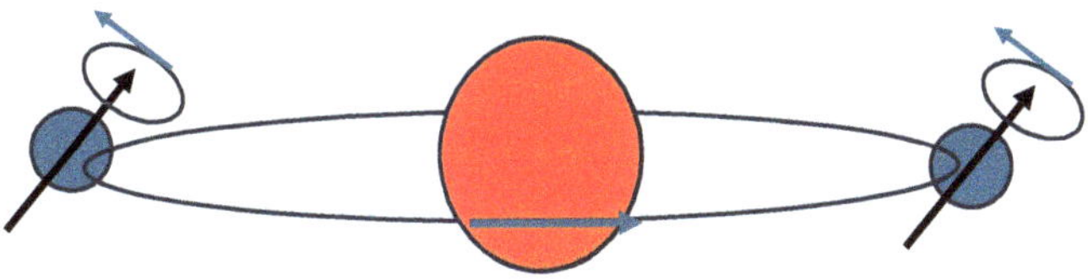

Figure 1.4 *Motion of the Earth with rotation and revolution around the Sun. The angle between the rotation axis and the revolution orbit makes the seasons on the Earth.*

The rotation of the Earth (from west to east) makes periodic sunrises and sunsets. The rotation periodicity is getting slower due to the gravitational interaction with the Moon. In 1.4 billion years ago, one day was 18 h. There is a short-term fluctuation of the rotational period because of the motion of the Earth's core. The rotation axis has an angle of 0.14π radian with the revolution orbit, which makes the periodic change of the seasons as shown in Figure 1.4. The period of the revolution is more stable than that of the rotation. The mechanism of the solar eclipse (God's curse at ancient dates) was clarified based on the Copernican theory.

The Copernican theory was established with the procedure shown above. The only remaining problem was that the annual parallax of fixed stars was not observed until the eighteenth century, because the measurement uncertainties of the direction of stars were larger than their seasonal variations. In 1838, Bessel observed the annual parallax of 61 Cygni with a value of 0.3 arcseconds, indicating the distance from the Earth to be 10.4 light-years (presently corrected to 11.4 light-years). The distances to the stars were clarified to be much larger than expected until the first observation of annual parallax.

The revolution of the Earth around the Sun was clarified by the accuracy improvement of clocks and the observation of phenomena outside the Earth.

1.5 The finite speed of light

Until the middle of the seventeenth century, the speed of light was believed to be either infinite or too fast to measure because no phenomenon was observed that highlighted the nonzero propagation time of light. However, the propagation time of light is not negligible for observing the phenomena on other planets using accurate clocks. It became clear that the measurement results of the revolution period of Io (one of Jupiter's satellites) were fluctuating. The fluctuation amplitude was larger than the measurement uncertainty. The variation of the period correlates with the variation of the size of Jupiter observed by Galilei, which indicates the

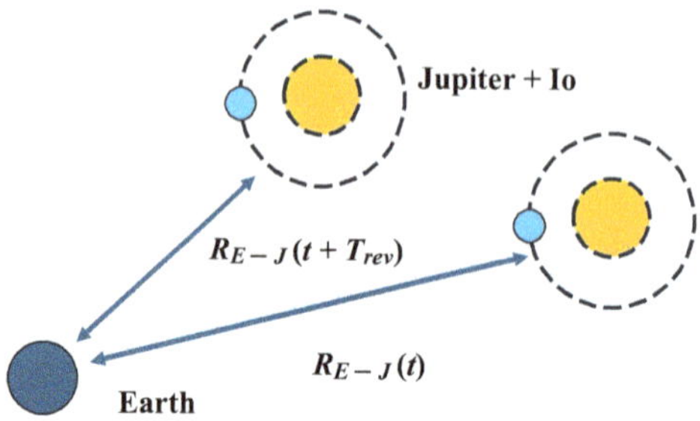

Figure 1.5 The observed period of the revolution of Io around Jupiter fluctuates because the distance between the Earth and Jupiter changes within the period and the propagation time of light also changes

periodic variation of the distance to Jupiter. In 1676, Rømer considered a finite propagation time of light between Jupiter and Earth. The period of the revolution is measured by $T_{rev} + [R_{E-J}(t + T_{rev}) - R_{E-J}(t)]/c$, where T_{rev} is the real revolution period, $R_{E-J}(t)$ is the distance between the Earth and Jupiter at the time of t, and c is the speed of light (see Figure 1.5). With this idea, the speed of light was estimated to be 2.2×10^8 m/s, which is 26% lower than the present value [4]. The irregular orbital period of Io was noted because of improvements in the accuracy of the time measurement. However, the accuracy of the distance between Earth and Jupiter was not high enough in that era.

The terrestrial measurement of the speed of light was performed by Fizeau and Foucault by measuring the propagation time to a mirror [5,6]. The measurement uncertainty was improved drastically because the distance to the mirror can be determined exactly. Fizeau determined the round-trip propagation time compared with the rotation velocity of the gear, with which the light from the gear recess is reflected by the mirror and blocked by the gear's convex. The speed of light was determined to be 3.13×10^8 m/s with this method. Foucault performed the measurement with the same method except for using a rotating mirror instead of the gear and obtained the result of 2.98×10^8 m/s, which is 0.6% different from the present value (299,792,458 m/s).

The speed of light became fundamental to electromagnetism because the electromagnetic field propagates with the speed of light (Chapter 3.3.3). The speed of light was clarified to be constant for any observers moving at different velocities, which is fundamental to the theory of relativity (Chapters 3.6 and 3.7).

1.6 The concept of atoms

In the fifth century BC, Democritus proposed the idea that all matters in the universe are made up of indivisible particles. This is the first to give the concept of atoms. He believed that atoms could not be broken down, although they can combine or separate. He proposed that the variations in the properties of different substances were due to the different arrangements of the atoms they contained [7]. However, there was no way to confirm this idea by experiment at that era. This idea has been forgotten for 2000 years.

Dalton revived the atomic theory, giving the following laws [8]:

1. Every element is constructed by atoms
2. Atom of an element is different from that of other elements, which are distinguished by the mass
3. Atoms of the same element have exactly the same properties
4. Compounds (molecules) are made by the binding of atoms
5. Chemical reactions change just the binding between atoms, but no atoms are produced or annihilated

These laws were established based on experimental results indicating that oxygen with a volume of V_O reacts with nitrogen with a volume of V_N or $2V_N$ (production of NO_2 and NO), but cannot react with the volume of nitrogen of $K_N V_N (1 < K_N < 2)$. Dalton listed six atoms: hydrogen (H), oxygen (O), nitrogen (N), carbon (C), sulfur (S), and phosphorus (P). Dolton tried to estimate the size of atoms.

Atoms were believed to be the unit that cannot be divided further until the end of the nineteenth century.

Here, we consider the estimation of the mass of gaseous atoms and molecules. The relation between pressure P and volume V is given by

$$PV = N_g k_B T \tag{1.1}$$

where N_g is the number of gaseous atoms or molecules, k_B is the Boltzmann constant, and T is the thermodynamic temperature. Before (1.1) was derived as shown in Chapter 3.4, Boyle's law (PV is constant with a constant T), Charles' law (PV is proportional to T), and Avogadro's law (N_g is constant for all kinds of gas with the same PV/T) were established. The mass ratio between different gaseous atoms or molecules is obtained by comparing the mass with the same PV/T.

The masses of atoms are approximately given by (atomic weight/Avogadro constant)/1000 kg, where atomic weight is the sum of the numbers of protons and neutrons. The Avogadro constant $N_A \left(= 6.02214076 \times 10^{23}\right)$ is the number of ^{12}C atoms (6 protons and 6 neutrons) in a carbon material of 12 g. The Avogadro constant was at first estimated by Q_E/e, where Q_E is the electric charge required to produce a material with the mass of (1 g × molecular weight) by electrolysis and e is the elementary charge (Chapter 1.7). Molecular weight is the sum of the atomic weights of constituent atoms. Currently, the Avogadro constant is obtained with an uncertainty of 10^{-9} with the method introduced in Chapter 2.5.

At the beginning of the twentieth century, there were still researchers who did not believe in the existence of atoms and molecules, but the final confirmation was given by Brownian motion (random motion of colloids in a liquid). Einstein analyzed the motion of colloids, giving a hypothesis that this random motion is caused by the collision of liquid molecules having random motion [9].

1.7 Atomic structure

In 1897, a beam of unknown negatively charged particles was discovered as a flow from cathode to anode (z-direction) [10]. The orbit of this beam was bent when an electric field E_e was applied in the perpendicular direction (x-direction); the orbit bends in the x-direction. When magnetic field B_m is applied in y-direction, the orbit bends to the x-direction. The electric charge/mass was obtained from the bending of the orbit by electric field. The velocity in the z-direction is obtained from the comparison between the electric field and magnetic field where the bending forces from both fields are balanced. The particle discovered with this experiment was called an "electron."

Millikan investigated that the electric charge of oil mist can only be integer multiples of the elementary charge $e\left(= 1.602176634 \times 10^{-19}\ \mathrm{C}\right)$ [11]. Compared with Thomson's experiment, the mass of the electron $m_e\left(= 9.1093837139 \times 10^{-31}\ \mathrm{kg}\right)$ was investigated to be much smaller than that of atoms.

It was clarified that atoms are not the smallest unit, but they have some structure, including electrons. There was a mystery about how electrons in atoms exist. Thomson proposed a "plum pudding model"; atoms are positively charged sphere, and electrons are distributed in them [12]. Nagaoka proposed a model that electrons revolve around a nucleus with a positive charge as shown in Figure 1.6 [13]. At that date, the Nagaoka model was not supported, because electrons revolving around the nucleus are believed to lose kinetic energy by emitting electromagnetic waves and finally merge with the nucleus (Chapter 3.3.4). With the Thomson model, atoms were expected not to give a strong electric force to charged particles. However, Rutherford's group observed a strong scattering of α-particles (He^{2+}) at the surface of gold, which indicated the strong repulsive Coulomb force from the nucleus [14]. The Nagaoka model was supported after this result was published, but the mystery of stable electron orbits remained until quantum mechanics was established. This was the first step to observe the phenomena in a microsize, for which Newtonian mechanics are not applicable.

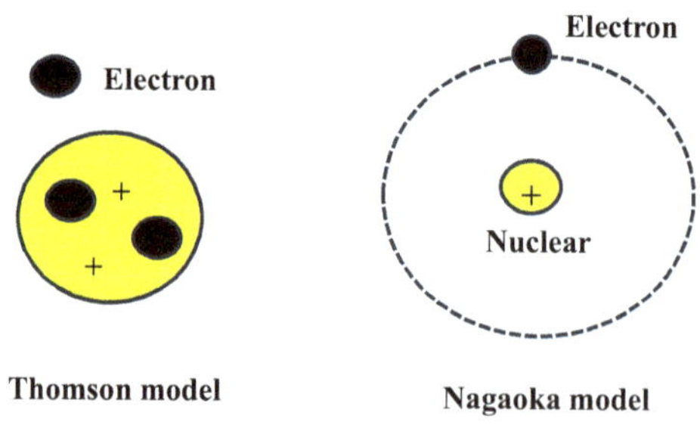

Figure 1.6 Atomic structure with the Thomson model and the Nagaoka model

1.8 Expansion of the observational domain in the twentieth century

This chapter introduces the procedures in science by expanding the observation area: large-distance areas and microsized. In the nineteenth century, spectroscopic analysis of light from stars started, which gave information on stellar structures [15]. Motion velocities are also obtained from the Doppler shift of the spectrum. Note that astronomical observation was limited to a small fraction of stars in the Milky Way. Observation of several galaxies outside the Milky Way was performed in the twentieth century (Chapter 6) [16]. Previously, quasars at 13.1 billion light-years distant were observed [17]. Expanding the observation area, much new astronomical information has been obtained.

The information for micro sizes also increased drastically in the twentieth century. Particles bound in a small area can take only discrete energy because particles have a wave nature and can take wavelengths resonant to the size of the area (Chapter 3.8.2). Atoms and molecules can take only discrete internal energies, and the state can be changed by absorbing or emitting photons (particle nature of electromagnetic waves) with the energy corresponding to the energy difference. The energy of a photon is proportional to the frequency; therefore, atoms and molecules absorb or emit electromagnetic waves with discrete frequencies. An atomic clock is a clock based on the frequencies of electromagnetic waves absorbed by atoms, which reduces the measurement uncertainty of time and frequency drastically (Chapter 4).

Expansion of the observed area is also required for energy. Until the nineteenth century, only temperature around room temperature was experienced. By obtaining ultra-high or low temperatures, many new results have been obtained. Until 1969, the proton was believed to be an elemental particle that cannot be divided further. But with a high-energy collision with an electron, it was proved to be a binding of two up quarks and one down quark (Chapter 9.2) [18]. Ultra-high energy is required to observe the phenomena inside nuclei because the energy gap between different binding energy states in a small area ($< 10^{-15}$ m) is very high. Since ultra-low temperature was obtained by producing liquid helium, superfluidity and super-conductivity were observed. A temperature below 10^{-3} K was attained by laser cooling, with which the wave natures of atoms become significant (Chapter 7).

References

[1] Torge W. and Müller J. (2012). *Geodesy*. De Gruyter Textbook; De Gruyter; 2012. p 5

[2] Swerdlow N. M. (2005). 'Ptolemy's theories of the latitude of the planets in the almagest, handy tables, and planetary hypotheses'. In: Buchwald, J. Z., Franklin, A. (eds.), *Wrong for the Right Reasons. Archimedes*, vol 11. Dordrecht, Springer. DOI:10.1007/1-4020-3048-7_3

[3] Gingerich O. 'The great Martian catastrophe and how Kepler fixed it'. *Phys. Today* 2011; 64 50–54. DOI:10.1063/PT.3.1259
[4] Filonovich S. R. (1986). *The Greatest Speed.* Moscow, Mir Publishers; 1986. p 285
[5] Fizeau H. 'Sur une expérience relative à la vitesse de propagation de la lumière'. *Comptes Rendus Hedomadaires des Seances de l'Academie des Science* 1849; 29 90–92
[6] Foucault L. 'Determination experimentale de la vitesse de la lumière'. *Comptes Rendus Hedomadaires des Seances de l'Academie des Science* 1852; 55 501–503
[7] Trefil J. and Bertsch G. F. https://www.britannica.com/science/atom/Development-of-atomic-theory; 2025
[8] Rocke A. J. 'In search of El Dorado: John Dalton and the origins of the atomic theory'. *Soc. Res.*, 2005; 72 1–34. DOI: 10.1353/sor.2005.0003
[9] Einstein A. 'Über die von der molekularkinetischen Theorie der Wärme geforderte Bewegung von in ruhenden Flüssigkeiten suspendierten Teilchen'. *Annahlen der Physik* 1905; 322 549–560. DOI:10.1002/andp.19053220806
[10] Thomson J. J. 'On the emission of negative corpuscles by the alkali metals'. *Philos. Mag. Ser. 6* 1905; 10 584–590. DOI:10.1080/14786440509463405
[11] Millikan R. A. 'On the elementary electrical charge and the Avogadro constant'. *Phys. Rev.*, 1913; 2 109–141. DOI: 10.1103/PhysRev.2.109
[12] Hon G. and Goldstein B. R. 'J.J. Thomson's plum-pudding atomic model: The making of a scientific myth'. *Annahlen der Physik* 2013; 525 A129–133. DOI:10.1002/andp.201300732
[13] Bryson B. *A Short History Nearly Everything.* Broadway Books; 2003
[14] Rutherford E. 'The scattering of α and β particles by matter and the structure of the atom'. *Lond. Edinb. Dubl. Phil. Mag. & J. Sci.,* 1911; 21 669–688. DOI:10.1080/14786440508637080
[15] Hearnshaw, J. B.*The Analysis of Starlight – One Hundred and Fifty Years of Astronomical Spectroscopy.* Cambridge University Press; 1986
[16] Hubble E. P. 'A spiral nebula as a stellar system, Messier 31'. *Astrophys. J.*, 1929; 69 103–158.
[17] Wang F., Yang J., Fan X. *et al.* 'A luminous quasar at redshift 7.642'. *Astrophys. J. Lett.*, 2021; 907 L1 1–7. DOI 10.3847/2041-8213/abd8c6
[18] Bloom E. D., Coward D. H., DeStaebler H. *et al.* 'High-energy inelastic $e{-}p$ scattering at 6° and 10°'. *Phys. Rev. Lett.*, 1969; 23 930–934. DOI:10.1103/PhysRevLett.23.930

Chapter 2

Measurement uncertainty and the establishment of standards and fundamental physical values

Abstract

The reduction of the measurement uncertainty is an important subject for physics research. This chapter introduces the procedure to establish the standard of time, length, and mass. The uncertainty of time and standards was reduced drastically by the development of the atomic clock. Defining the speed of light and the Planck constant, length, and mass are measured via the measurement of time and frequency.

2.1 Introduction

Precision measurement of physical values is important to discover slight effects that have previously never been observed. Mass (M), length (L), and time (T) are fundamental values of physics, and all physical values are given by the form $M^{n_M} L^{n_L} T^{n_T}$ and (n_M, n_L, n_T) for velocity, acceleration, and energy are $(0, 1, -1)$, $(0, 1, -2)$, and $(1, 2, -2)$, respectively. Therefore, attainable measurement uncertainty of some physical value is given by those for these three values.

In the SI unit, the units of mass, length, and time are given by kilogram (kg), meter (m), and second (s), respectively. This chapter introduces the standards of these three values.

2.2 Measurement uncertainty

Performing the measurement, there is always a question about its reliability. The reliability of measured values can be confirmed by repeating the measurements many times. The measurement results are expected to be distributed within a certain limited range. The uncertainty given by the nonzero distribution area is called the "statistical uncertainty."

The measurement results depend on the circumstances. Real values are defined as the measurement with specified conditions. It is not realistic to measure with the ideal condition; therefore, there is always some shift from the defined real value.

The uncertainty of the shift from the defined value is "systematic uncertainty." If we know the dependence of the measurement on the circumstance, we can give a correction and the systematic uncertainty is reduced. The systematic uncertainty cannot be zero because of the uncertainties of corrected values.

2.2.1 Statistical uncertainty

The reliability of the measurement result is confirmed by repeating the measurement and confirming the reproducibility. The obtained results are always distributed over a nonzero area, which gives a statistical uncertainty. The distribution area of the measurement result is sometimes given by the uncertainty principle (a fundamental of quantum mechanics), which prohibits the perfect deterministic physical values (Chapter 3.8.3).

The central value in the distribution is estimated by averaging the measurement results. Increasing the number of measurement results, the averaged value becomes closer to the central value in the distribution area, and the statistical uncertainty is reduced. Here, we assume that the measurement of a value X is obtained by two results $X_c \pm \sigma_s$ with the probabilities of 1/2. From the total number of measurement results N_s, results of $X_c + \sigma_s$ and $X_c - \sigma_s$ are obtained with the numbers of n_s and $N_s - n_s$, respectively. The average of the measurement results is

$$X_{ave} = X_c + \sigma_s \left[\frac{2n_s}{N_s} - 1 \right] \tag{2.1}$$

with the probability of

$$P_s(n_s) = \frac{N_s!}{n_s!(N_s - n_s)} \left(\frac{1}{2} \right)^{N_s}. \tag{2.2}$$

X_{ave} is obtained within the area of

$$X_c - \frac{2\Delta n_s}{N_s} \sigma_s < X_{ave} < X_c + \frac{2\Delta n_s}{N_s} \sigma_s, \tag{2.3}$$

where Δn_s is the standard deviation of n_s obtained by

$$(\Delta n_s)^2 = \sum \left(n_s - \frac{N_s}{2} \right)^2 P_s(n_s) = \sum n_s^2 P_s(n_s) - \frac{N_s^2}{4}$$

$$\sum n_s^2 P_s(n_s) = \sum n_s(n_s - 1) P_s(n_s) + \sum n_s P_s(n_s)$$

$$\sum n_s(n_s - 1) P_s(n_s) = \frac{N_s(N_s - 1)}{4} \sum \frac{(N_s - 2)!}{(n_s - 2)!(N_s - n_s)} \left(\frac{1}{2} \right)^{N_s - 2} = \frac{N_s(N_s - 1)}{4}$$

$$\sum n_s P_s(n_s) = \frac{N_s}{2} \sum \frac{(N_s - 1)!}{(n_s - 1)!(N_s - n_s)} \left(\frac{1}{2} \right)^{N_s - 1} = \frac{N_s}{2}$$

$$\Delta n_s = \frac{\sqrt{N_s}}{2} \tag{2.4}$$

and

$$|X_c - X_{ave}| < \frac{\sigma_s}{\sqrt{N_s}} \tag{2.5}$$

is derived. Equation (2.5) indicates that the statistical uncertainty is reduced by averaging many measurement samples, as shown in Figure 2.1.

The central limit theorem indicates the possibility to take the average value of X_{ave} is given by

$$P(X_{ave}) = \frac{\sqrt{N_s}}{\sigma_s\sqrt{\pi}}\exp\left[-N_s\left(\frac{X_{ave} - X_c}{\sigma_s}\right)^2\right] \tag{2.6}$$

with any distribution of measurement results [1].

Here, we consider the fluctuation of the flow of particles with the average rate of N_c/s. The temporal current is at a minimum of zero and the distribution area is approximately given by $\sigma_s = N_c$/s. The statistical uncertainty is given by $\sigma_{st} = \sqrt{N_c}$ /s, which is called "shot noise." The current fluctuates between the values of $N_c\left(1 - \frac{1}{\sqrt{N_c}}\right)$ and $N_c\left(1 + \frac{1}{\sqrt{N_c}}\right)$ as shown in Figure 2.2. The effect of shot noise is suppressed as the flow rate becomes higher. Shot noise is observed with the electric current (current of charged particles) or light intensity (current of photons) (see Chapter 3.5.2).

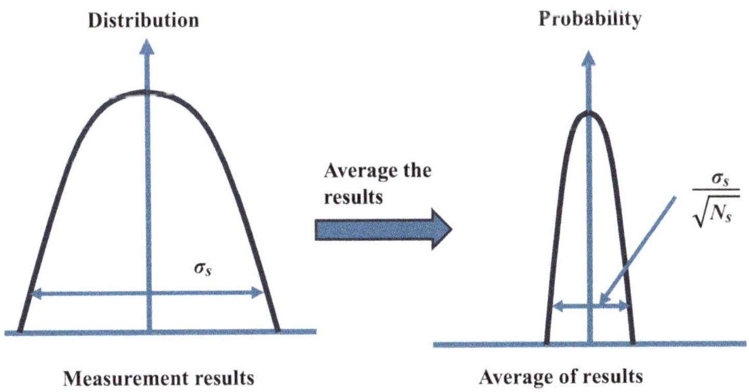

Figure 2.1 Concept of statistical uncertainty. Measurement results are distributed within a finite area σ_s. Averaging the results, the uncertainty is reduced by a factor of $1/\sqrt{N_s}$, where N_s is the number of measurement samples.

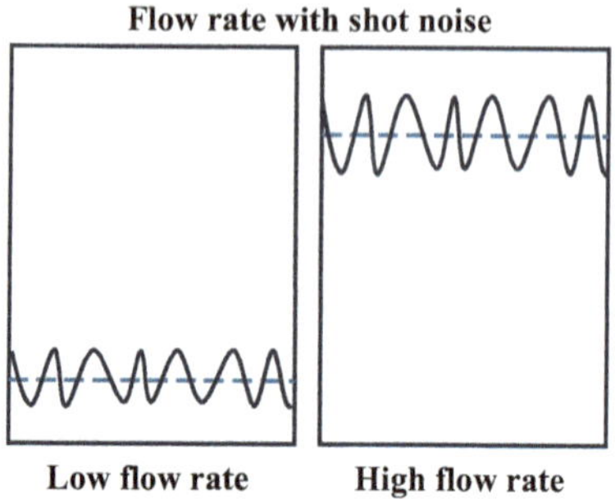

Figure 2.2 *Flow rate with shot noise. With a high flow rate, the influence of shot noise is suppressed.*

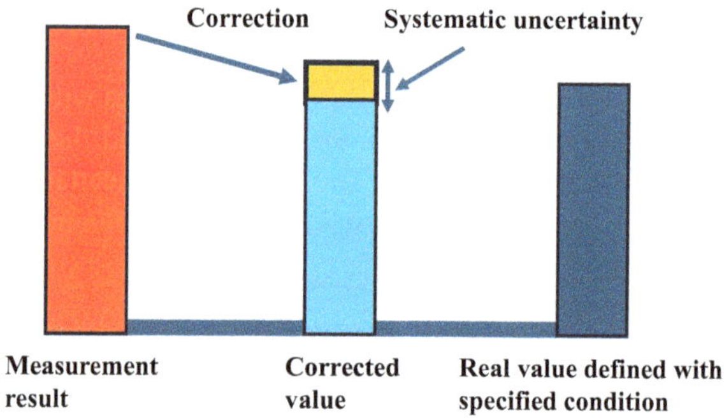

Figure 2.3 *The real value is defined by the measurement with specified conditions. The actual measurements are shifted from the real value, and a correction can be given. The uncertainty in the correction is systematic uncertainty.*

2.2.2 Systematic uncertainty

The measurement results might be distributed in an area shifted from the real value because the measurement values can vary according to a given set of circumstances. The standard values should be defined under a specified condition, and the measured value is shifted under another circumstance. The uncertainty of the shift of measurement from the defined value is the "systematic uncertainty." If we know the dependence of the measurements on the circumstances, we can correct the measured values using the estimated shift. For example, the real length should be defined with a specified temperature because there is thermal expansion. The length at a defined temperature is estimated by monitoring the temperature and giving a correction with a known expansion coefficient. The systematic uncertainty is reduced by this treatment; however, it is not zero because of the uncertainties of corrected values (uncertainties of temperature and expansion coefficient) as shown in Figure 2.3.

2.2.3 Total uncertainty

As shown above, the measurement uncertainty is given by statistical uncertainty Δ_{st} and systematic uncertainty Δ_{sy}. The total uncertainty is given by

$$\Delta_{tot} = \sqrt{\Delta_{st}^2 + \Delta_{sy}^2}. \tag{2.7}$$

The systematic uncertainty is given by the shifts induced by different causes. From the uncertainties from each cause Δ_{sy-k}, the total systematic uncertainty is given by

$$\Delta_{sy} = \sqrt{\sum_k \Delta_{sy-k}^2}. \tag{2.8}$$

2.3 Time standard

Time is the value that can be measured with periodic phenomena. In ancient times, the periodicity was observed with the change between day and night. At first, the time of 1 h was defined by dividing the interval time between sunrise and sunset by twelve with the idea of a duodecimal system. However, there was a problem that the length of the day depends on the season. This definition was changed to the interval between sunrises being 24 h (86,400 s). After the Copernican theory was established, this time interval was recognized as the period of the rotation of the Earth. However, the period of the rotation was clarified to not be constant, as shown in Chapter 1.4. It can also fluctuate over time scales of 10–50 years by a factor of 10^{-8} [2].

During the period 1956–1967, the standard of time was changed to the period of Earth's orbit around the Sun because it was clarified to be more constant than the period of the Earth's rotation. However, the fluctuation of the period is caused by the gravitational perturbations from other planets. Note also that the Sun is losing mass, and hence, the radius of Earth's orbit about the Sun is increasing over a long period. Therefore, the orbital period is getting longer. Its fluctuation is of order 10^{-9}, which is just one order smaller than that for the rotation period of the Earth [2].

After World War II, atomic clocks were invented based on the following quantum mechanical concepts (detailed in Chapter 3.8).

1. Each electromagnetic wave also has a particle nature (called the photon) having energy $h\nu$ (h: Planck constant, ν: frequency) and momentum of h/λ (λ: wavelength) in the propagation direction.
2. Not only photons, but also all particles have the nature of waves, having the frequency and wavelength indicating the energy and momentum, respectively. All particles (for example, electrons in an atom) bound in a limited area can take only discrete energy because the wavelength should be resonant with the size of the area.

3. The electrons in atoms can change their state (make a transition) with the energy gap of ΔE absorbing or emitting a photon satisfying $\Delta E = h\nu$ so that the total energy is conserved. Therefore, atoms absorb or emit electromagnetic waves with discrete frequencies.

An atomic clock is based on the frequency of the electromagnetic waves (e.g., light, microwave) absorbed or emitted by atoms (Chapter 4.1). The measurement uncertainty of time and frequency was drastically reduced since the development of atomic clocks because the atomic transition frequencies are discrete values, which are uniquely determined by the Coulomb interaction between electrons and nuclei (bonding force). Since the development of atomic clocks, time and frequency are the physical values that can be measured with uncertainties much lower than any other physical values.

Measuring uncertainty cannot be zero, also for atomic clocks, for the following reasons. There is a nonzero systematic uncertainty because the measurement is performed in a circumstance where the electromagnetic perturbation is not zero. Atomic transitions with large ratios of bonding force/electromagnetic perturbation are advantageous to measure with low systematic uncertainty, as shown in Figure 2.4. Use of atoms with low velocity is preferable to reduce the quadratic Doppler shift (frequency shift induced by the atomic motion as a relativistic effect shown in Chapters 3.6 and 5.1). The measurement of the transition frequency has a slight dependence on the gravitational potential (Chapters 3.7 and 5.1). The definition of ideal transition frequency is the measurement with (i) zero electromagnetic noise, (ii) zero motion velocity, and (iii) measurement on the geoid surface. Systematic uncertainty can be reduced by giving corrections to the measurement results by monitoring the electromagnetic field, atomic motion velocity, and altitude of the measurement place. There is also statistical uncertainty because

Figure 2.4 The time and frequency measured by the atomic clock have finite uncertainty because of the electromagnetic perturbation and the spectrum linewidth. Atomic transitions with large ratios of (bonding force/electromagnetic perturbation) and (transition frequency/linewidth) are advantageous to be used for atomic clocks.

the transition is induced when the frequency of the electromagnetic wave is shifted from the transition frequency within a linewidth given by the uncertainty principle between energy and measurement time, as shown in Figure 2.4 (Chapters 3.8.3, 4.1, and 4.3). Transitions with a high ratio of transition frequency/spectrum linewidth are advantageous to reduce statistical uncertainty. Statistical uncertainty can be reduced by averaging many measurement samples. Using many atoms, low statistical uncertainty is obtained with a short measurement time.

With the commercial atomic clock using a microwave transition of a Cs atom, the measurement uncertainty is the order of 10^{-13} (takes 300,000 years to make the error of 1 s). The uncertainty of 10^{-16} was obtained with a laboratory-type Cs atomic clock using laser-cooled atoms (Chapter 4.2.4). Since 1967, the transition frequency of cesium (Cs) atoms in the microwave region has been defined to be 9,192,631,770 Hz [2,3]. The use of optical transition frequencies is more advantageous than microwave transition frequencies for precision measurement because the time of 1 s is measured with a finer scale. Since the measurement of frequency of laser light became possible (Chapter 4.2.5), researchers were eager to develop atomic clocks based on optical transition frequencies. Recently, the measurement uncertainty of 10^{-18} was obtained with several atomic transition frequencies in the optical region (Chapter 4.3) [4–7], which can be used as a novel frequency standard with the uncertainty lower than that for the Cs transition frequency. But we need more discussion about which transition frequency is most appropriate as the standard. The conclusion might be given after 2030.

Now, other physical values are also measured via the measurement of time and frequency, defining proper constants as shown below, because of the ultra-low measurement uncertainty of time and frequency.

2.4 Length standard

The concept of length was also clear in ancient times, as it was easy to compare the lengths of different bars. Soon after the French Revolution, the length of one meter was advised to be defined as one ten-millionth of the distance between the North Pole and the equator. In 1799, a standard bar of platinum was made based on the meridional meter length, called "Mètre de Archives." In 1869, the definition of the meter was changed to this "Mètre de Archives" because it was difficult to measure the size of the Earth. There was the uncertainty of order 10^{-5} given by the bending and thermal expansion.

In 1889, the definition of a meter was changed to the distance between two lines on a standard bar measured at the melting point of ice to minimize the effect of thermal expansion. To minimize the bending, the standard bar was made of an alloy of platinum with 10% iridium (the alloy is harder than pure platinum) with an X cross-section (to minimize the bending). The uncertainty given by the thermal expansion, bending, and the thickness (1 μm) of the lines is of order 10^{-7}.

In the 1940s, the use of the wavelength of light emitted by an atom was proposed as the length standard, because the properties of the atoms in the gaseous

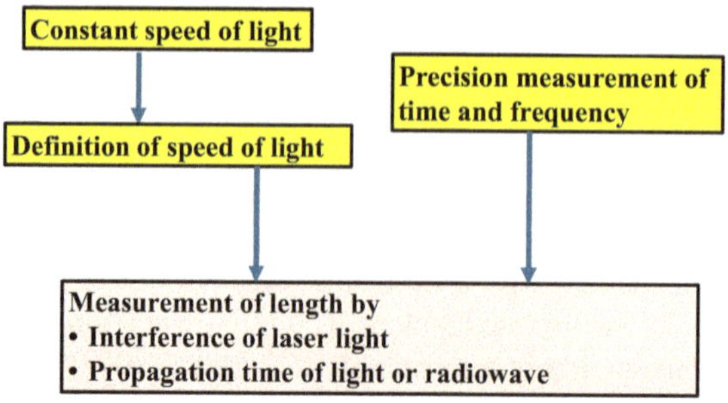

Figure 2.5 *Precision measurement of length using the defined speed of light and an atomic clock*

state are universal. In 1960, a new definition of the meter length was given in terms of the wavelength ($6.05780211 \times 10^{-7}$ m) of fluorescence light emitted by krypton-86 (^{86}Kr); the uncertainty was reduced to 4×10^{-9} [8]. However, the frequency bandwidth of this light is as large as a few MHz. Further reduction of measurement uncertainty was difficult until laser light with a narrow frequency bandwidth was developed.

The speed of light (also microwave, radio wave, etc.) in a vacuum was confirmed to be constant for any observers moving with different velocities (Chapter 3.5.1) [9,10]. The speed of light in the vacuum c was defined to be 299,792,458 m/s [11]. With the constant speed of light, the uncertainty of wavelength is the same as that of frequency.

Now several atomic transition frequencies were measured with the uncertainty of 10^{-18} (Chapter 4.3). Short distance measurement can be performed in comparison with the wavelength of laser light, observing the interference between irradiated and reflected lights as shown in Figure 2.5. Long distance is often measured with the round-trip (by reflection) propagation time of pulsed laser light. The distance to satellites or planetary probes is measured by the propagation time of radio waves. An accurate clock is required to measure the distance with low uncertainty. Note also that the speed of light or radio waves in the air is slightly shifted from that in a vacuum, which was defined.

2.5 Mass standard

The base unit of mass, 1 kg, was originally defined as the mass of distilled water in a cubic container of size 10 cm × 10 cm × 10 cm (i.e., 1 L of water). But it could not be a good standard to reproduce the same mass, because it is difficult to obtain ideal distilled water. The accuracy of the container size is also questionable. The water density has a dependence on the temperature; therefore, it was defined as the temperature of 4 °C, where the density is highest. The accuracy of the temperature was also questionable. Therefore, the uncertainty of the mass could not be low enough to be a standard mass.

From 1889 to 2019, the kg standard was defined by an artifact, the "International Prototype of the Kilogram (IPK)" [12,13]. IPK is stored at the International Bureau of Weights and Measures (BIPM) on the outskirts of Paris. It is a right-circular cylinder of 39 mm made from the alloy of platinum (90%) and iridium (10%). The hardness of pure platinum is enhanced by the addition of iridium to minimize the scratches on the surface. Moreover, the right-circular shape was used to minimize the change of the mass by the attachment of dust, which is proportional to the surface area. Six copies of the IPK were also made, but there is a mass difference from that of the IPK with the order of 10^{-8} kg [14] because of limits in the accuracy of size and mixture rates of iridium during fabrication. In addition, mass differences with copies have drifted over time. The masses of IPK have increased with dust in the air settling on its surfaces and have also diminished from the original values when it was cleaned. The mass uncertainty is of order 5×10^{-8}.

The theory of relativity indicates the relation between the energy E and mass M as (Chapter 3.6)

$$E = Mc^2 \tag{2.9}$$

Considering the relation between energy and frequency $E = h\nu$, (2.9) is rewritten as

$$M = \frac{h\nu}{c^2} \tag{2.10}$$

In 2019, the Planck constant h was defined to be $6.62607015 \times 10^{-34}$ J/Hz [15]. After the definition of h and c, the attainable measurement uncertainty of mass can be the same as that for time and frequency in principle. The mass of an electron m_e is determined with the uncertainty of 10^{-10} from the precision measurement of the transition frequencies of the H atom, as shown in Figure 2.6.

The mass ratio between atoms and electrons is measured from the ratio of the frequencies of cyclotron motions (circular motion under the magnetic field) of X^{j+}

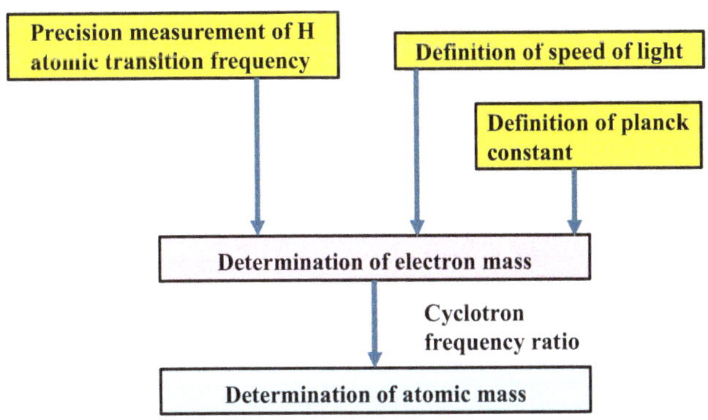

Figure 2.6 Determination of atomic mass after the definition of the speed of light and the Planck constant

ion (mass $m(X^{j+})$) and electron (mass m_e). For example, $m_e/m(^{12}C^{6+})$ was obtained to be $4.572753097 \times 10^{-5}$. Giving a correction to this value, taking the mass of lacking electrons and the binding energy into account, $m_e/m(^{12}C) = 4.57149926 \times 10^{-5}$ was obtained [15]. With this method, the atomic mass can be measured with the uncertainty of 10^{-10} as shown in Figure 2.6.

The mass standard in the macro size can also be defined with higher accuracy if we can estimate the number of atoms. A mass of 1 kg can be obtained from a 93.6 mm-diameter sphere of silicon (^{28}Si) atoms. ^{28}Si was chosen because ultra-pure defect-free monocrystalline silicon already exists in fabricated form. The natural abundance of the ^{28}Si isotope is 0.922, which can be improved up to 0.99999 by isotope separation [16]. The lattice defect is minimized by using a sphere crystal (with a cube, it is significant at the end). The lattice spacing between atoms in the crystal, $d = 0.192$ nm, can be measured by observing the focused X-ray irradiated on the crystal. Passing and blocking of X-rays are observed periodically by shifting the position of the crystal. The shifted position is monitored by laser interferometry, as shown in Figure 2.7. The size of this Si sphere is measured using optical interferometry and the volume was measured with an uncertainty of 1.8×10^{-8}. With this method, the mass uncertainty of the Si sphere is expected to be reduced more than that of the kg standard. This measurement of the size of the Si sphere and the lattice spacing should be performed at the same temperature. The temperature should be stabilized within 1 mK. The uncertainty of the mass is currently 2.1×10^{-8} [17].

Substance quantity is the parameter to give the number of atoms or molecules in macro size, whose uncertainty is equivalent to that of mass. It is given with the unit of mol from (number of atoms and molecules/N_A), where N_A is Avogadro's constant. The N_A is defined as "number of ^{12}C atoms contained in a carbon of 12 g (0.012 kg)." The recommended value of N_A in 2014 is $6.022140857(74) \times 10^{23}$ [18]. The mass of atoms or molecules is roughly obtained by (number of protons + neutrons)/(1000 N_A) with the unit of kg. This estimation is not strictly accurate because of (1) a slight difference in mass between proton and neutron and (2) lack of mass due to the binding energy.

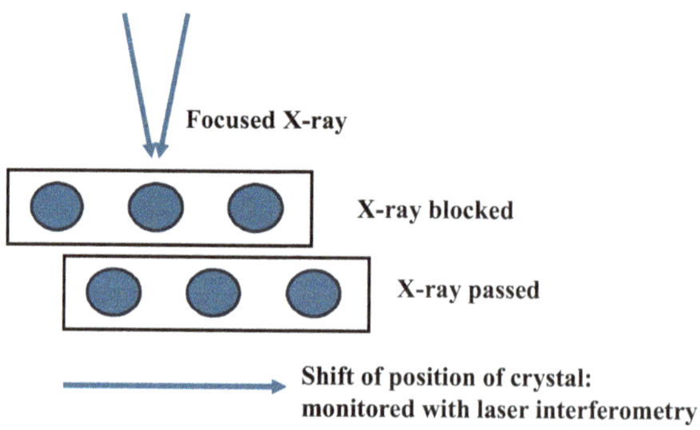

Figure 2.7 Measurement of the spacing between atoms in a crystal

The mol electron mass is given by $A_e = m_e/m(^{12}C) \times 0.012$ kg, and the Avogadro constant is given by

$$N_A = \frac{A_e}{m_e} \tag{2.11}$$

The electron mass was determined after the definition of the Planck constant, and the Avogadro constant is determined with an uncertainty below 10^{-9}.

2.6 Other physical standards

2.6.1 Temperature

The thermodynamic temperature T is a parameter to show the broadening of the energy distribution. With a thermal equilibrium state, the distribution at each state with energy E is proportional to $\Omega \exp(-E/k_B T)$ as shown in Chapter 3.4. Ω is the number of states with energy of E and k_B is the Boltzmann constant. Until 2019, the thermodynamic temperature was defined by taking the triple point of distilled water with a defined abundance ratio of isotopes to be 273.16 K, which is determined with an uncertainty of 0.1 mK. It is not realistic to use water as defined, and the uncertainty in the actual temperature measurement is higher. The Boltzmann constant k_B has been measured with the uncertainty below 10^{-6} through the measurement of the velocity distribution of gaseous atoms or molecules with the temperature stabilized to the triple point of distilled water [11]. Velocity distribution is monitored from the broadening of the atomic or molecular spectrum given by the Doppler effect or the acoustic velocity.

In 2019, the Boltzmann constant was defined to be $k_B = 1.380649 \times 10^{-23}$ J/K [11]. With this new definition, the thermodynamic temperature is a parameter of energy distribution. For example, the temperature of an object can be determined from the precise measurement of the peak frequency of the blackbody radiation (Chapter 3.5).

2.6.2 Electric current

The understanding of electromagnetic theory has been summarized by Maxwell's equations (Chapter 3.3.2). Electric current is a flow of electric charge, and the unit of ampere (A = C/s) is used. However, a measurement of the number of flowing charged particles per unit time was not realistic. So, the electric current of 1 A has been defined as the "constant current that produces a force of attraction of 2×10^{-7} Newtons (kg m s^{-2}) per meter of length between two straight parallel conductors of infinite length and negligible cross-section placed one meter apart in vacuum." The conductor in reality must have a finite length and a nonzero cross-section. The accuracy of the attractive force is also limited. Therefore, the electric current cannot be measured with low uncertainty while this definition is used. But ultra-low uncertainty of electric current was not required for the research of physics and engineering until the twentieth century.

Recently, microchip electric circuits with small electric currents were developed, and the electric current can be determined from the number of flowing charged particles (for example, electrons). In 2019, the elementary charge was defined to be $e = 1.602176634 \times 10^{-19}$ C [11]. With this definition, electric current can be measured as the product of the elementary charge and the flow of the charged particle. The definition of the elementary charge makes a great contribution to physics because the electric charges of all particles are integer multiples of the elementary charge.

2.6.3 Luminous intensity

Luminous intensity is a parameter to show the brightness of a light source, which human eyes perceive. The power of irradiated light is given by the unit of W (=J/s = kg m^2 s^{-3}). But the sensitivity of human eyes to the light depends on the color (frequency ν). Luminous intensity is a measure of the frequency-weighted power in a direction per unit solid angle. The unit of luminous intensity is candela (cd), and one candela was derived from the brightness of typical candles. With the SI unit, one candela is defined as "brightness of light with a frequency of 540 THz (wavelength 555 nm) with a radiant intensity if 1/683 W per steradian." A factor to give the dependence on the luminous intensity on the frequency is given by a function $L_m(\nu)$, which is maximum at 540 THz and zero in the infrared and ultraviolet regions [19].

There is no change in the definition of the luminous intensity, even after the definitions of the Planck constant, elementary charge, and Boltzmann constant. But defining the Planck constant, the energy of one photon is determined with the uncertainty of the frequency. Measuring the flow of photons, we can determine the luminous intensity with the uncertainty of the frequency because $L_m(\nu)$ is an artificial factor.

2.7 From macro to micro standards

The standards for physical quantities were originally defined using macro-sized artifacts (solid substances) and have recently been changed to micro-size objects (gaseous atoms). It is difficult to define physical values from a solid object because we cannot make an exact prototype of the standard because there are limits to the accuracy of its size and purity of substance. In contrast, the properties of atoms or molecules in a gaseous state are generally stable and universal.

The advantage of time and frequency is that the value of microscopic value (atomic or molecular transition frequency) can be measured with a microwave frequency counter. Now, also optical frequency can be measured using an optical frequency comb (Chapter 4.2.5). After the speed of light was defined as a constant value, the wavelength of light became a scale of length with the uncertainty of the frequency (10^{-18} for some laser lights). Long distance is often measured from the propagation time of light or radio waves. However, note that the speed of light in the air is slightly shifted from that in a vacuum, which was defined.

Determinations of the mass of particles in constructing atoms or molecules (electron, proton, neutron, etc.) are possible from the precision measurement of atomic or molecular transition frequencies. For the mass measurement of objects with macrosize, we need the accurate number of atoms, whose uncertainty has been reduced to 10^{-8}. Laser interferometry makes a great contribution to measuring the number of atoms.

References

[1] Feller W. *An Introduction to Probability Theory and Its Applications. I* (third ed.). John Wiley & Sons, Inc.; 1968

[2] *Bereau International des Poids et Mesures.* The International System of Units; 2019. p 207

[3] Vanier J. and Audoin C. *The Quantum Physics Atomic Frequency Standards.* Bristol and Philadelphia, Adam, Hilger; 1989. p 610

[4] Chou C. W., Hume D. B., Koelemeij J. C. J., Wineland D. J. and Rosenband T. 'Frequency comparison of two high-accuracy Al^+ optical clocks'. *Phys. Rev. Lett.*, 2010; 104 70802. DOI:10.1103/PhysRevLett.104.070802

[5] Huntemann N., Sanner C., Lipphardt B., Tamm Chr. and Peik E. 'Single-ion atomic clock with 3×10^{-18} systematic uncertainty'. *Phys. Rev. Lett.*, 2016; 116 063001. DOI:10.1103/PhysRevLett.116.063001

[6] Ushijima I., Takamoto M., Das M., Ohkubo T. and Katori H. 'Cryogenic optical lattice clocks'. *Nat. Photon.*, 2015; 9 185–189. DOI:10.1038/nphoton. 2015.5

[7] Nicholson T. L., Campbell S. L., Hutson R. B. *et al.* 'Systematic evaluation of an atomic clock at 2×10^{-18} total uncertainty'. *Nat. Comm.*, 2015; 6 6896. DOI:10.1038/ncomms7896

[8] Phelps F. M. '^{86}Kr vacuum wavelengths'. *J. Opt. Soc. Am.*, 1964; 54 864–868. DOI:10.1364/JOSA.54.000864

[9] Michelson A. A. 'The relative motion of the earth and of the luminiferous ether'. *Am. J. Sci.*, 1881; 22 120–129. DOI:10.2475/ajs.s3-22.128.120

[10] Brillet A. and Hall J. L. 'Improved laser test of the isotropy of space'. *Phys. Rev. Lett.*, 1979; 42 549–552. DOI:10.1103/PhysRevLett.42.549

[11] *Bereau International des Poids et Mesures.* The International System of Units. 2019; p 199

[12] Davis R. 'The SI unit of mass'. *Metrologia* 2003; 40 299–305. DOI:10.1088/0026-1394/40/6/001

[13] *Bereau International des Poids et Mesures.* The International System of Units. 2019; p 208

[14] Jabbour Z. J. and Yaniv S. L. 'The kilogram and measurements of mass and force'. *J. Res. Nat. Inst. Stand. Tech.*, 2001; 106 25–46. DOI:10.6028/jres.106.003

[15] Farnham D. L. *et al.* 'Determination of the electron's atomic mass and the proton/electron mass ratio via penning trap mass spectroscopy'. *Phys. Rev. Lett.*, 1995; 75 3598–3601. DOI:10.1103/PhysRevLett.75.3598

[16] Abrosimov N. V. *et al.* 'A new generation of 99.999% enriched 28Si single crystals for the determination of Avogadro's constant'. *Metrologia* 2017; 54 599. DOI:10.1088/1681-7575/aa7a62

[17] Kuramoto N. *et al.* 'Realization of the new kilogram by the XRCD method using ^{28}Si-enriched spheres'. *Meas. Sens.*, 2021; 18 100091. DOI:10.1016/j.measen.2021.100091

[18] *Bereau International des Poids et Mesures.* The International System of Units. 2019; p 209

[19] *Bereau International des Poids et Mesures.* The International System of Units. 2019; p 210

Chapter 3

Fundamentals of classical and early modern physics

Abstract

This chapter introduces the fundamentals of classical mechanics, electromagnetism, statistical mechanics, mysteries of light (constancy of speed of light, dual nature as wave and particle), the theory of relativity, quantum mechanics, and relativistic quantum mechanics. This knowledge is required to understand the contents of Chapters 4–9.

3.1 Introduction

This book introduces the procedure of the development of physics by the expansion of the observation area and the reduction of measurement uncertainty. However, knowledge of fundamental physics is required to understand the contents in the following chapters. This chapter summarizes the fundamentals of physics, minimizing the equations so that it is easy to read, even for beginners.

3.2 Classical mechanics

3.2.1 Newtonian mechanics

Newtonian mechanics is based on the following three laws.

1. An object continues in uniform straight-line motion with the velocity of $\vec{v}$ as long as no force acts on it.
2. The sum of the forces $\vec{F}$ on an object is equal to the product of the mass m and the acceleration $\vec{a} = (d\vec{v}/dt)$,

$$\vec{F} = m\vec{a} = m\frac{d\vec{v}}{dt} \tag{3.1}$$

We can define the mass as a parameter to indicate the difficulty of giving the acceleration. Defining momentum $\vec{p} = m\vec{v}$ (this definition is not valid with the theory of relativity and quantum mechanics), (3.1) is rewritten as

$$\vec{F} = \frac{d\vec{p}}{dt} \tag{3.2}$$

3. When one body exerts a force on a second body, the second body simultaneously exerts a force equal in magnitude and opposite in direction on the first body.

$$\overrightarrow{F_{1,2}} = -\overrightarrow{F_{2,1}} \tag{3.3}$$

Considering the interaction between two bodies 1 and 2,

$$\frac{d\overrightarrow{p_1}}{dt} = \overrightarrow{F_{1,2}} \quad \frac{d\overrightarrow{p_2}}{dt} = \overrightarrow{F_{2,1}} = -\overrightarrow{F_{1,2}}$$

$$\frac{d\left[\overrightarrow{p_1} + \overrightarrow{p_2}\right]}{dt} = 0, \tag{3.4}$$

and the conservation of the total momentum (valid also with modern physics) is derived.

The "Work" is defined as

$$\Delta W = \int \vec{F} \cdot d\vec{r} \tag{3.5}$$

where $\vec{r}$ is the position vector ($\vec{v} = d\vec{r}/dt$). The kinetic energy (work to accelerate a body from zero velocity) is given by

$$K_E = \int \vec{F} \cdot d\vec{r} = \int m \frac{d\vec{v}}{dt} \cdot \vec{v} dt = \frac{m[\vec{v}]^2}{2} = \frac{|\vec{p}|^2}{2m}. \tag{3.6}$$

The change of potential energy (ability of work) is given by

$$P_E = -\int \vec{F} \cdot d\vec{r}, \tag{3.7}$$

The temporal change of the total energy $E = K_E + P_E$ is given by

$$\frac{dP_E}{dt} = \frac{\partial P_E}{\partial t} + \frac{\partial P_E}{\partial \vec{r}} \frac{d\vec{r}}{dt}$$

$$\frac{dK_E}{dt} = \int \vec{F} \cdot \frac{d\vec{r}}{dt} = -\frac{\partial P_E}{\partial \vec{r}} \frac{d\vec{r}}{dt}$$

$$\frac{dE}{dt} = \frac{\partial P_E}{\partial t}. \tag{3.8}$$

Total energy is conserved without the temporal change of the potential energy at a given position $P_E(\vec{r})$. The nonzero $(\partial P_E(\vec{r})/\partial t)$ is given by a work from outside.

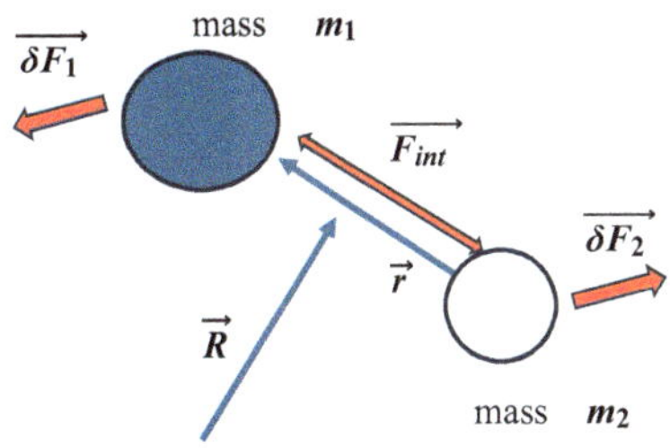

Figure 3.1 Diagram showing the forces acting on the two bodies

Here, we consider two-body motion (masses: m_1 and m_2) based on the motion of the center of mass and the relative motion, as shown in Figure 3.1. Bodies 1 and 2 experience the external forces $\overrightarrow{F_1}$ $(= \overrightarrow{\delta F_1} + \overrightarrow{F_{\text{int}}})$ and $\overrightarrow{F_2}(= \overrightarrow{\delta F_2} - \overrightarrow{F_{\text{int}}})$, respectively. Here, $\pm\overrightarrow{F_{\text{int}}}$ is the interaction force between the two bodies.

$$m_1 \frac{d^2 \overrightarrow{r_1}}{dt^2} = \overrightarrow{F_1} \tag{3.9}$$

$$m_2 \frac{d^2 \overrightarrow{r_2}}{dt^2} = \overrightarrow{F_2} \tag{3.10}$$

(3.9) + (3.10)

$$(m_1 + m_2)\frac{d^2 \vec{R}}{dt^2} = \overrightarrow{F_1} + \overrightarrow{F_2} = \overrightarrow{\delta F_1} + \overrightarrow{\delta F_2}$$

$$\vec{R} = \frac{m_1 \overrightarrow{r_1} + m_2 \overrightarrow{r_2}}{m_1 + m_2} \tag{3.11}$$

$$(3.9) \times \frac{m_2}{m_1 + m_2} - (3.10) \times \frac{m_1}{m_1 + m_2}$$

$$\mu_r \frac{d^2 \vec{r}}{dt^2} = \frac{m_2}{m_1 + m_2}\overrightarrow{F_1} - \frac{m_1}{m_1 + m_2}\overrightarrow{F_2} = \overrightarrow{F_{\text{int}}} + \overrightarrow{F_r}$$

$$\overrightarrow{F_r} = \frac{m_2}{m_1 + m_2}\overrightarrow{\delta F_1} - \frac{m_1}{m_1 + m_2}\overrightarrow{\delta F_2}$$

$$\vec{r} = \overrightarrow{r_1} - \overrightarrow{r_2}, \ \mu_r = \frac{m_1 m_2}{m_1 + m_2} \tag{3.12}$$

$\vec{R}$ denotes the position of the center of mass, and $\vec{r}$ is the relative position. μ_r is the reduced mass. Equations (3.11) and (3.12) show the motion of the center of mass and the relative motion between two bodies, respectively. Here we consider the motion that both bodies are combined and the force $\delta\vec{F_1}$ is given to body 1. To keep the distance between two bodies constant,

$$\overrightarrow{F_{\text{int}}} = -\frac{m_2}{m_1 + m_2}\overrightarrow{\delta F_1}$$

$$\frac{d^2\overrightarrow{r_1}}{dt^2} = \frac{d^2\overrightarrow{r_2}}{dt^2} = \frac{1}{m_1 + m_2}\overrightarrow{\delta F_1} \tag{3.13}$$

is derived. $\overrightarrow{F_r}$ can include the component perpendicular to $\vec{r}$, which is the force that changes the rotational angular velocity. No change in the rotational angular velocity was induced when $\overrightarrow{\delta F_2} = (m_2/m_1)\overrightarrow{\delta F_1}$. The change of the rotational angular velocity is often discussed using the angular momentum given by $\vec{L} = \vec{r} \times \vec{p}$ ($\vec{p} = \mu_r(d\vec{r}/dt)$) as follows:

$$\frac{d\vec{L}}{dt} = \vec{v} \times \vec{p} + \vec{r} \times \frac{d\vec{p}}{dt} = \vec{r} \times \overrightarrow{F_r}\,(\text{called "torque"}) \tag{3.14}$$

The angular momentum is conserved ($d\vec{L}/dt = 0$) when the potential energy depends only to $\vec{r}$ $\left(\overrightarrow{F_r} \| \vec{r}\right)$. Kepler's second law (Chapter 1.4) indicates the conservation of angular momentum. Using the two-dimensional polar coordinate $(x, y) = (r\cos\theta, r\sin\theta)$, the angular momentum is $L = \mu_r r^2(d\theta/dt)$. With constant velocity,

$$\mu_r\frac{d^2r}{dt^2} = \mu_r\frac{d}{dt}\left(\frac{dr}{dt}\right) = \mu_r\frac{d}{dt}\left[\frac{dr}{dx}\frac{dx}{dt} + \frac{dr}{dy}\frac{dy}{dt}\right] = \mu_r\frac{d}{dt}\left[\cos\theta\frac{dx}{dt} + \sin\theta\frac{dy}{dt}\right]$$

$$= \mu_r\left[-\frac{dx}{dt}\sin\theta\frac{d\theta}{dt} + \frac{dy}{dt}\cos\theta\frac{d\theta}{dt} + \cos\theta\frac{d^2x}{dt^2} + \sin\theta\frac{d^2y}{dt^2}\right] = \mu_r r\left(\frac{d\theta}{dt}\right)^2$$

$$\left(\frac{d^2x}{dt^2} = \frac{d^2y}{dt^2} = 0, \quad \frac{dx}{dt} = \frac{dr}{dt}\cos\theta - r\frac{d\theta}{dt}\sin\theta, \quad \frac{dy}{dt} = \frac{dr}{dt}\sin\theta + r\frac{d\theta}{dt}\cos\theta\right) \tag{3.15}$$

is derived as the "centrifugal force," which is an apparent force in the radial direction.

Newton presented the law of universal gravitation (published in 1687), showing the gravitational attractive force F_G between two bodies with mass of $m_{1,2}$ takes the form,

$$F_G = G\frac{m_1 m_2}{r^2}, \tag{3.16}$$

where G is the gravitational constant. The proportionality between the mass and the gravitational force was derived from the experimental fact that gravitational acceleration does not depend on the mass. Kepler's third law (Chapter 1.4) indicates that $(d\theta/dt)^2\, r^3$ is constant, which is consistent with the balance between the centrifugal force given by (3.15) and the gravitational force given by (3.16). The law of universal gravitation unified the attractive force between planets and stars and the force that makes objects fall to the ground. Moreover, the acceleration due to gravity on the surface of the Earth is almost constant because the height of fall near the Earth's surface is much less than the radius of the Earth. Only motions of macroscopic objects were observed at that era, which were explained by Newtonian mechanics. The validity of (3.16) has been confirmed for a distance larger than 1 mm. However, it is still questionable if this relation is also valid for the micro size.

3.2.2 Equations of motion in an electromagnetic field

A moving object with a velocity $\vec{v}$ and an electric charge of q_e experiences a force due to an electric field $\vec{E_e}$ and a magnetic field $\vec{B_m}$ as follows.

$$\vec{F} = q_e\left[\overrightarrow{E_e} + \vec{v} \times \overrightarrow{B_m}\right]$$
$$F_x = q_e\left[E_{ex} + v_y B_{mz} - v_z B_{my}\right]$$
$$F_y = q_e\left[E_{ey} + v_z B_{mx} - v_x B_{mz}\right] \quad (3.17)$$
$$F_z = q_e\left[E_{ez} + v_x B_{my} - v_y B_{mx}\right]$$

Here, we consider the electromagnetic field using the voltage Φ_e and vector potential $\vec{A}$

$$\overrightarrow{E_e} = -\nabla\Phi_e - \frac{\partial\vec{A}}{\partial t}\left(E_{eq} = -\frac{\partial\Phi_e}{\partial q} - \frac{\partial A_q}{\partial t} \quad q = x, y, z\right)$$
$$\vec{B_m} = \nabla \times \vec{A}\left(B_{mx} = \frac{\partial A_z}{\partial y} - \frac{\partial A_y}{\partial z}, B_{my} = \frac{\partial A_x}{\partial z} - \frac{\partial A_z}{\partial x}, B_{mz} = \frac{\partial A_y}{\partial x} - \frac{\partial A_x}{\partial y}\right) \quad (3.18)$$

Taking $\overrightarrow{B_m} = (0, 0, B_m)$, (3.17) is rewritten as

$$F_x = q_e\left(E_{ex} + v_y B_m\right)$$
$$F_y = q_e\left(E_{ye} - v_x B_m\right). \quad (3.19)$$
$$F_z = q_e E_{ez}$$

Using momentum, (3.19) is rewritten as

$$\frac{dp_x}{dt} = q_e E_{ex} + q_e v_y B_m = -q_e\frac{d\Phi_e}{dx} - q_e\frac{dA_x}{dt} + q_e v_y\frac{\partial A_y}{\partial x}$$
$$\frac{dp_y}{dt} = q_e E_{ey} - q_e v_x B_m = -q_e\frac{d\Phi_e}{dy} - q_e\frac{dA_y}{dt} + q_e v_x\frac{\partial A_x}{\partial y}$$
$$\frac{dp_z}{dt} = -q_e\frac{d\Phi_e}{dz}$$
$$\frac{d\left(\vec{p} + q_e\vec{A}\right)}{dt} = \frac{q_e}{m}\frac{\partial\vec{p}\cdot\vec{A}}{\partial\vec{r}} - q_e\frac{d\Phi_e}{d\vec{r}} \quad (3.20)$$

Lagrange equation given by

$$L_G = K - P_E = \frac{m}{2}\left|\frac{d\vec{r}}{dt}\right|^2 - P_E = \frac{|\vec{p}|^2}{2m} - P_E$$
$$\frac{d}{dt}\left[\frac{dL_G}{d(d\vec{r}/dt)}\right] = \frac{\partial L_G}{\partial \vec{r}} \tag{3.21}$$

is useful to solve motion equations with any coordinates. Using (3.21),

$$\frac{d\vec{p}}{dt} = -\frac{\partial P_E}{\partial \vec{r}} \tag{3.22}$$

is derived. Equation (3.20) is derived using the Lagrange equation with the following transformation:

$$\vec{p} \to \vec{p} + q_e\vec{A}$$
$$K = \frac{\left|\vec{p} + q_e\vec{A}\right|^2}{2m} \approx \frac{|\vec{p}|^2}{2m} + \frac{\vec{p}\cdot\vec{A}}{m} \tag{3.23}$$
$$P_E \to P_E - q_e\Phi_e.$$

This relation is also valid to discuss the electromagnetic field with the theory of relativity and quantum mechanics. The kinetic energy is transformed to

$$\frac{\left(\vec{p} + q_e\vec{A}\right)\cdot\left(\vec{p} + q_e\vec{A}\right)}{2m} = \frac{|\vec{p}|^2}{2m} + \frac{q_e}{m}\vec{p}\cdot\vec{A} \tag{3.24}$$

Taking $\vec{A} = (-yB_m, xB_m, 0)/2$, the energy shift by the magnetic field (called Zeeman shift) is given by

$$\frac{q_e}{m}\vec{p}\cdot\vec{A} = \frac{q_e}{2m}B_m\left[xp_y - yp_x\right] = \frac{q_e}{2m}(\vec{r}\times\vec{p})\cdot\overrightarrow{B_m} = \frac{q_e}{2m}\vec{L}\cdot\overrightarrow{B_m} \tag{3.25}$$

3.3 Light as an electromagnetic wave

3.3.1 Discussion up to the establishment of the electromagnetic wave

The identity of light had been a mystery until the nineteenth century. The phenomena like reflection by a mirror, diffraction when passing through a slit, and interference between two lights were observed with light. In the seventeenth century, "particle theory" and "wave theory" were proposed. The reflection is possible for both particles and waves. The observation of diffraction seemed to indicate the wave nature of light. The observation of the interference also made the wave theory advantageous, although it was not reproducible. The wave theory was not decisively advantageous, because nobody could explain what kind of oscillation propagates in what kind of medium. The identity as an electromagnetic wave is derived as follows.

3.3.2 *Derivation of the Maxwell's equations*

In the eighteenth and nineteenth centuries, several laws of the electric field $\overrightarrow{E_e}$ and the magnetic field $\overrightarrow{B_m}$ were established. In 1864, these laws were unified in a set of equations known as Maxwell's equations, shown below (in SI units) [1].

$\nabla \cdot \overrightarrow{E_e} = \frac{\rho_e}{\varepsilon}$ (ρ_e: the electric charge density, ϵ: the permittivity)

$$\frac{\partial E_{ex}}{\partial x} + \frac{\partial E_{ye}}{\partial y} + \frac{\partial E_{ez}}{\partial z} = \frac{\rho_e}{\varepsilon} \tag{3.26}$$

$$\nabla \times \overrightarrow{E_e} = -\frac{\partial \overrightarrow{B_m}}{\partial t}$$

$$\frac{\partial E_{ez}}{\partial y} - \frac{\partial E_{ey}}{\partial z} = -\frac{\partial B_{mx}}{\partial t}, \quad \frac{\partial E_{ex}}{\partial z} - \frac{\partial E_{ez}}{\partial x} = -\frac{\partial B_{my}}{\partial t}, \quad \frac{\partial E_{ey}}{\partial x} - \frac{\partial E_{ex}}{\partial y} = -\frac{\partial B_{mz}}{\partial t}, \tag{3.27}$$

$$\nabla \cdot \overrightarrow{B_m} = 0$$

$$\frac{\partial B_{mx}}{\partial x} + \frac{\partial B_{my}}{\partial y} + \frac{\partial B_{mz}}{\partial z} = 0, \tag{3.28}$$

$$\nabla \times \overrightarrow{B_m} = \mu\left[\vec{j} + \varepsilon\frac{\partial \overrightarrow{E_e}}{\partial t}\right]$$

($\vec{j}$: the electric current density, μ: the permeability)

$$\frac{\partial B_{mz}}{\partial y} - \frac{\partial B_{my}}{\partial z} - \mu\left[j_x \mid c\frac{\partial E_{ex}}{\partial te}\right], \quad \frac{\partial B_{mx}}{\partial z} - \frac{\partial B_{mz}}{\partial x} = \mu\left[j_y + \varepsilon\frac{\partial F_{cy}}{\partial t}\right],$$

$$\frac{\partial B_{my}}{\partial x} - \frac{\partial B_{mx}}{\partial y} = \mu\left[j_z + \varepsilon\frac{\partial E_{ez}}{\partial t}\right]. \tag{3.29}$$

Equation (3.26) is derived from Coulomb's law, indicating that the electric field from a point electric charge q_e is given by

$$\overrightarrow{E_e} = \frac{q_e}{4\pi\varepsilon r^3}\vec{r}. \tag{3.30}$$

Using the polar coordinate ($x = r\sin\theta\cos\varphi$, $y = r\sin\theta\sin\varphi$, $z = r\cos\theta$)

$$\iint \overrightarrow{E_e} \cdot d\vec{S} = \iint \left|\overrightarrow{E_e}\right| r^2 \sin\theta \, d\theta d\varphi = \frac{q_e}{\varepsilon}$$

$$\left|\overrightarrow{E_e}\right| = \int \frac{\partial}{\partial r}\left|\overrightarrow{E_e}\right| dr$$

$$\iint \left|\overrightarrow{E_e}\right| r^2 \sin\theta d\theta d\varphi = \iiint \left[\frac{\partial \left|\overrightarrow{E_e}\right|}{\partial r}\right] dV = \frac{1}{\varepsilon} \iiint \rho_e, \; dV \; dV = r^2 \sin\theta dr d\theta d\phi$$

$$\frac{\partial}{\partial r} = \frac{\partial x}{\partial r}\frac{\partial}{\partial x} + \frac{\partial y}{\partial r}\frac{\partial}{\partial y} + \frac{\partial z}{\partial r}\frac{\partial}{\partial z} = \sin\theta\cos\phi\frac{\partial}{\partial x} + \sin\theta\sin\phi\frac{\partial}{\partial y} + \cos\theta\frac{\partial}{\partial z}$$

$$E_{ex} = \sin\theta\cos\phi\left|\vec{E}\right|, \; E_{ey} = \sin\theta\sin\phi\left|\vec{E}\right|, \; E_{ez} = \cos\theta\left|\vec{E}\right|$$

$$\frac{\partial E_{ex}}{\partial x} + \frac{\partial E_{ye}}{\partial y} + \frac{\partial E_{ez}}{\partial z} = \frac{\rho_e}{\varepsilon}. \tag{3.31}$$

and (3.26) is derived. Since there is no magnetic charge, (3.28) is derived from (3.26) by changing $\overrightarrow{E_e} \to \overrightarrow{B_m}, \rho_e \to 0$.

Stokes's theorem for the line integral of a vector quantity (obtained as a vector value) is required for the derivations of (3.27) and (3.29). Here, we consider the line integral in the xy-plane from which the vector component in the z-direction is obtained. The line integral of the closed loop of a vector quantity $\vec{f}$ is given by

$$\oint \vec{f} \cdot d\overrightarrow{l_z} = \oint f_x dx + \oint f_y dy. \tag{3.32}$$

Stokes's theorem indicates

$$\oint \vec{f} \cdot d\overrightarrow{l_z} = \iint_{inside\ loop} \left[\frac{\partial f_y}{\partial x} - \frac{\partial f_x}{\partial y}\right] dS_z, \; dS_z = dxdy, \tag{3.33}$$

which is derived for the vector quantity in the tangential direction of the circular loop, $\vec{f} = (f_x, f_y) = f(r)(-\sin\theta, \cos\theta)$ as follows.

$$\frac{\partial}{\partial x} = \frac{\partial r}{\partial x}\frac{\partial}{\partial r} + \frac{\partial \tan\theta}{\partial x}\frac{\partial\theta}{\partial \tan\theta}\frac{\partial}{\partial\theta} = \cos\theta\frac{\partial}{\partial r} - \frac{\sin\theta}{r}\frac{\partial}{\partial\theta},$$

$$\frac{\partial}{\partial y} = \sin\theta\frac{\partial}{\partial r} + \frac{\cos\theta}{r}\frac{\partial}{\partial\theta},$$

$$\iint \left[\frac{\partial f_y}{\partial x} - \frac{\partial f_x}{dy}\right] d\overrightarrow{S_z} = \int_0^{2\pi} d\theta \int_0^r \left[\frac{\partial f(r)}{\partial r} + \frac{f(r)}{r}\right] r dr = 2\pi r f(r) = \oint \vec{f} \cdot d\overrightarrow{l_z} \tag{3.34}$$

Faraday's law of induction shows that the voltage in the electric loop is given by

$$\oint \overrightarrow{E_e} d\vec{l} = \iint \left[\nabla \times \overrightarrow{E_e}\right] \cdot d\vec{S} = \iint_{inside\ loop} \frac{\partial \overrightarrow{B_m}}{\partial t} \cdot d\vec{S} \tag{3.35}$$

and (3.27) is derived. Ampere's circuital law states that:

$$\oint \overrightarrow{B_m} \cdot d\vec{l} = \iint \left[\nabla \times \overrightarrow{B_m}\right] \cdot d\vec{S} = \iint_{inside\ loop} \mu\vec{j} \cdot d\vec{S}. \tag{3.36}$$

However, there is a discrepancy in the case of a circuit containing a condenser for this law. From (3.36), there should not be a magnetic field around the condenser. This discrepancy was solved using the constancy of the total electric charge:

$$\nabla \cdot \vec{j} + \frac{\partial \rho_e}{\partial t} = \nabla \cdot \left[\vec{j} + \varepsilon \frac{\partial \overrightarrow{E_e}}{\partial t}\right] = 0. \tag{3.37}$$

Equation (3.36) was corrected to

$$\oint \overrightarrow{B_m} \cdot d\vec{l} = \iint \left[\nabla \times \overrightarrow{B_m}\right] \cdot d\vec{S} = \iint_{inside\ loop} \mu \left[\vec{j} + \varepsilon \frac{\partial \overrightarrow{E_e}}{\partial t}\right] \cdot d\vec{S} \tag{3.38}$$

and (3.29) is derived. This correction made a large contribution to the establishment of the idea of electromagnetic waves, as shown in Chapter 3.3.3.

3.3.3 Derivation of electromagnetic wave

Maxwell's equation made a revolution in electromagnetism because the formula of electromagnetic waves is derived as follows. Taking the AC electric field direction in the x-direction, (3.27) gives

$$\frac{\partial E_{ex}}{\partial z} = -\frac{\partial B_{my}}{\partial t}, \quad \frac{\partial E_{xe}}{\partial y} = \frac{\partial B_{mz}}{\partial t}, \tag{3.39}$$

and (3.29) with $\vec{j} = 0$ is expressed as:

$$\frac{\partial B_{mz}}{\partial y} - \frac{\partial B_{my}}{\partial z} = \varepsilon\mu \frac{\partial E_{ex}}{\partial t}. \tag{3.40}$$

The AC electric field in the x-direction induces an AC magnetic field in the y- or z-direction; here we take the AC magnetic field in the y-direction. Thus, we have

$$\frac{\partial E_{ex}}{\partial z} = -\frac{\partial B_{my}}{\partial t} \rightarrow \frac{\partial^2 E_{ex}}{\partial z^2} = -\frac{\partial^2 B_{my}}{\partial z \partial t},$$

$$\frac{\partial B_{my}}{\partial z} = -\varepsilon\mu \frac{\partial E_{ex}}{\partial t} \rightarrow \frac{\partial^2 B_{my}}{\partial z \partial t} = -\varepsilon\mu \frac{\partial^2 E_{ex}}{\partial t^2},$$

$$\frac{\partial^2 E_{ex}}{\partial z^2} = \varepsilon\mu \frac{\partial^2 E_{ex}}{\partial t^2}. \tag{3.41}$$

To solve (3.41), we assume that E_{ex} is the product of the function of t, $T(t)$, and that of z, $Z(z)$.

$$\frac{1}{T}\frac{\partial^2 T}{\partial t^2} = \frac{1}{\varepsilon\mu Z}\frac{\partial^2 Z}{\partial z^2} = -(2\pi\nu)^2 \text{ (constant)},$$

$$E_{ex} = T(t)Z(z) = E_{e0}\sin\left[2\pi\nu\left(t \pm \sqrt{\varepsilon\mu}z\right) + \phi_0\right]$$

$$B_{my} = B_{m0}\sin\left[2\pi\nu\left(t \pm \sqrt{\varepsilon\mu}z\right) + \varphi_0\right], \; B_{m0} = -\sqrt{\varepsilon\mu}E_{e0}. \tag{3.42}$$

The solutions of E_{ex} and B_{my} are obtained as the formula of a wave propagating in the $\pm z$-directions. A change in the electric field induces a change in the magnetic field, and vice versa. As shown above, the directions of the electric field, the magnetic field, and the propagation direction are orthogonal to each other. The estimated propagation speed $c_m = 1/\sqrt{\varepsilon\mu}$ corresponds to the speed of light, as measured by Fizeau and Foucault [2,3]. The difference in the speed of light in the water and in the air was also measured [4]. Therefore, the light is proven to be an electromagnetic wave. It was also confirmed that radio and microwaves were also electromagnetic waves with frequencies lower than light. X-rays and γ-rays are also electromagnetic waves with higher frequencies than light. For the electromagnetic wave propagating in the z-direction, the direction of the electric field (magnetic field) can be in the x- or y-direction. Linear polarization indicates that the direction of the electric field (magnetic field) is constant, and circular polarization indicates that the direction of the electric field (magnetic field) rotates in the right (σ^+) or left (σ^-) direction. The interference of two lights is observed only when the polarization of both lights is equal. The light intensity (energy per unit time) is given by

$$I_{light} = c_m S\left[\langle P_{el}\rangle_{ave} + \langle P_{mg}\rangle_{ave}\right]$$

$$\langle P_{el}\rangle_{ave} = \frac{\varepsilon E_{e0}^2}{4}, \; \langle P_{mg}\rangle_{ave} = \frac{B_{m0}^2}{4\mu}, \quad S\text{: area of the light spot.} \tag{3.43}$$

Here, P_{el} and P_{mg} indicate the energy densities of electric and magnetic fields, respectively. From (3.42), $\langle P_{el}\rangle_{ave} = P_{mg_{ave}}$ is obtained. When the light is reflected by a mirror, the electric field on the mirror is zero, and the phase of the electric field in the reflected light changes by π ($E_{e0} \rightarrow -E_{e0}$). The phase of the magnetic field does not change by the reflection ($B_{m0} \rightarrow B_{m0}$). Taking $z = 0$ at the mirror position, the standing wave formed by the interference between the incident and reflected waves is given by

$$E_{ex} = 2E_{e0}\cos\left[2\pi\nu t + \varphi_0\right]\sin\left[2\pi\nu\sqrt{\varepsilon\mu}z\right]$$

$$B_{my} = 2B_{m0}\sin\left[2\pi\nu t + \varphi_0\right]\cos\left[2\pi\nu\sqrt{\varepsilon\mu}z\right]. \tag{3.44}$$

When the light is reflected by a mirror, the mirror gets the radiation pressure as shown by Figure 3.2, which is the Lorentz force induced by the electric current on

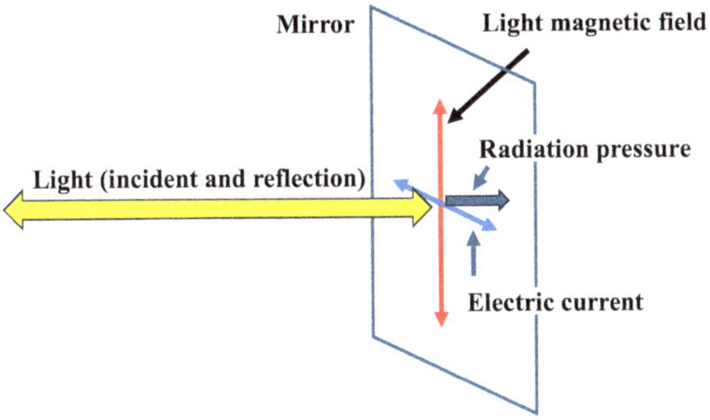

Figure 3.2 *Radiation pressure on the mirror reflecting a light is the Lorentz force between the electric current on the mirror and the light's magnetic field*

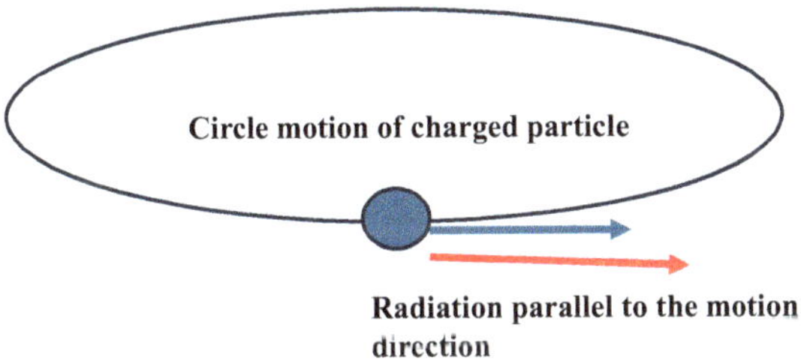

Figure 3.3 *Radiation induced by the circular motion of a charged particle*

the mirror surface ($j_x = B_y/\mu$ from (3.29)) and the magnetic field of B_{my} given by

$$F_z^{rad} = Sj_x B_{y_{ave}} = S\frac{B_y^2}{\mu}_{ave} = \frac{2}{c_m} I_{light}. \tag{3.45}$$

We can consider that the light has a momentum p_{light} $\left(= I_{light}/c_m\right)$ in the direction of propagation. The change of the momentum $p_{light} \to -p_{light}$ is caused by reflection. The change of momentum per unit time is the radiation pressure.

3.3.4 Radiation by the circular motion of charged particles

The electromagnetic field is induced by the change in the electric field distribution. Here, we consider the radiation of the electromagnetic wave by the circular motion of a charged particle shown in Figure 3.3. The important point is that the radiation term reduces to zero with the limit of $c_m \to \infty$. This result is reasonable to see that the change of the magnetic field is not induced with

$\sqrt{\varepsilon\mu} = 1/c_m \to 0$, as shown in (3.42), and the propagation of the electromagnetic wave could not exist.

Describing Maxwell's equation using the voltage and the vector potential,

$$\nabla^2\Phi_e + \nabla\cdot\frac{\partial\vec{A}}{\partial t} = -\frac{\rho_e}{\varepsilon}$$

$$\nabla\times\nabla\times\vec{A} = \mu\left[\vec{j} - \varepsilon\left(\frac{\partial^2\vec{A}}{\partial t^2} + \frac{\partial(\nabla\Phi_e)}{\partial t}\right)\right]$$

$$\text{using } \nabla\times\nabla\times\vec{A} = \nabla\left(\nabla\cdot\vec{A}\right) - \nabla^2\vec{A}$$

$$\left[\nabla^2 - \varepsilon\mu\frac{\partial^2}{\partial t^2}\right]\vec{A} - \nabla\left(\nabla\cdot\vec{A} + \varepsilon\mu\frac{\partial\Phi_e}{\partial t}\right) = -\mu\vec{j}. \tag{3.46}$$

The solutions of $\vec{A}$ and Φ_e to give determined $\overrightarrow{E_e}$ and $\overrightarrow{B_m}$ are not uniquely determined. Using an appropriate gauge (called the Lorenz gauge) satisfying:

$$\nabla\cdot\vec{A} + \varepsilon\mu\frac{\partial\Phi_e}{\partial t} = 0, \tag{3.47}$$

Equation (3.46) can be simplified to:

$$\nabla^2\Phi_e - \varepsilon\mu\frac{\partial^2\Phi_e}{\partial t^2} = -\frac{\rho_e}{\varepsilon}$$

$$\nabla^2\vec{A} - \varepsilon\mu\frac{\partial^2 A}{\partial t^2} = -\mu\vec{j}, \tag{3.48}$$

When $\rho_e = 0$ and $\vec{j} = 0$, (3.48) is the equation that describes wave propagation. Considering the propagation time,

$$\Phi_e(\vec{r}, t) = \frac{1}{4\pi\varepsilon}\iiint\frac{\rho_e\left(\vec{r'}, t - \frac{|\vec{r}-\vec{r}'|}{c_m}\right)}{|\vec{r}-\vec{r}'|}dV' \tag{3.49}$$

$$\vec{A}(\vec{r}, t) = \frac{\mu}{4\pi}\iiint\frac{\vec{j}\left(\vec{r'}, t - \frac{|\vec{r}-\vec{r}'|}{c_m}\right)}{|\vec{r}-\vec{r}'|}dV'. \tag{3.50}$$

Here, we consider the distribution of the electromagnetic field when a particle with an electric charge of q_e has a rotational motion $(x, y, z) = (r_0\cos(\omega t), r_0\sin(\omega t), 0)$, which results in a current of $q_e(-\omega r_0\sin(\omega t), \omega r_0\cos(\omega t), 0)$.

Assuming $|\vec{r}| \gg r_0$, the components of $\vec{A}$ are given by

$$A_x(\vec{r},t) = \frac{\mu}{4\pi}\frac{-q_e\omega r_0 \sin\left[\omega\left(t - \frac{|\vec{r}|}{c_m}\right)\right]}{|\vec{r}|}$$

$$A_y(\vec{r},t) = \frac{\mu}{4\pi}\frac{q_e\omega r_0 \cos\left[\omega\left(t - \frac{|\vec{r}|}{c_m}\right)\right]}{|\vec{r}|}$$

$$A_z(\vec{r},t) = 0. \tag{3.51}$$

From (3.47),

$$\frac{\partial \Phi_e(\vec{r},t)}{\partial t} = -\frac{1}{\varepsilon\mu}\left[\frac{\partial A_x}{\partial x} + \frac{\partial A_y}{\partial y} + \frac{\partial A_z}{\partial z}\right]$$

$$= \frac{q_e r_0}{4\pi\varepsilon}\left[\frac{\omega\left\{-x\sin\left[\omega\left(t - \frac{|\vec{r}|}{c_m}\right)\right] + y\cos\left[\omega\left(t - \frac{|\vec{r}|}{c_m}\right)\right]\right\}}{|r|^3}\right.$$

$$\left. + \frac{\omega^2\left\{-x\cos\left[\omega\left(t - \frac{|\vec{r}|}{c_m}\right)\right] - y\sin\left[\omega\left(t - \frac{|\vec{r}|}{c_m}\right)\right]\right\}}{|r|^2 c_m}\right] \tag{3.52}$$

is obtained and

$$\Phi_e(\vec{r},t) = \frac{q_e r_0}{4\pi\varepsilon}\left[\frac{\left\{x\cos\left[\omega\left(t - \frac{|\vec{r}|}{c_m}\right)\right] + y\sin\left[\omega\left(t - \frac{|\vec{r}|}{c_m}\right)\right]\right\}}{|r|^3}\right.$$

$$\left. + \frac{\omega\left\{-x\sin\left[\omega\left(t - \frac{|\vec{r}|}{c_m}\right)\right] + y\cos\left[\omega\left(t - \frac{|\vec{r}|}{c_m}\right)\right]\right\}}{|r|^2 c_m} + \frac{1}{|\vec{r}| r_0}\right] \tag{3.53}$$

is given. The third term of (3.53) is the time-independent term, which gives the DC voltage. At a long distance, the first term is negligibly small in comparison with the second term. The magnetic field is given by

$$B_{mx} = \frac{\partial A_z}{\partial y} - \frac{\partial A_y}{\partial z} = \frac{\mu}{4\pi}\frac{q_e\omega r_0 z\cos\left[\omega\left(t - \frac{|\vec{r}|}{c_m}\right)\right]}{|\vec{r}|^3} + \frac{\mu}{4\pi}\frac{q_e\omega^2 r_0 z\sin\left[\omega\left(t - \frac{|\vec{r}|}{c_m}\right)\right]}{|\vec{r}|^2 c_m}$$

$$B_{my} = \frac{\partial A_x}{\partial z} - \frac{\partial A_z}{\partial x} = \frac{\mu}{4\pi}\frac{q_e\omega r_0 z\sin\left[\omega\left(t - \frac{|\vec{r}|}{c_m}\right)\right]}{|\vec{r}|^3} + \frac{\mu}{4\pi}\frac{q_e\omega^2 r_0 z\cos\left[\omega\left(t - \frac{|\vec{r}|}{c_m}\right)\right]}{|\vec{r}|^2 c_m}$$

$$B_{mz} = \frac{\partial A_y}{\partial x} - \frac{\partial A_x}{\partial y}$$

$$= \frac{\mu}{4\pi} \frac{q_e \omega r_0 \left\{ x\cos\left[\omega\left(t - \frac{|\vec{r}|}{c_m}\right)\right] + y\sin\left[\omega\left(t - \frac{|\vec{r}|}{c_m}\right)\right]\right\}}{|\vec{r}|^3}$$

$$+ \frac{\mu}{4\pi} \frac{q_e \omega^2 r_0 \left\{ x\sin\left[\omega\left(t - \frac{|\vec{r}|}{c_m}\right)\right] - y\cos\left[\omega\left(t - \frac{|\vec{r}|}{c_m}\right)\right]\right\}}{|\vec{r}|^2 c_m}. \quad (3.54)$$

The circular motion of charged particles induces radiation with a frequency of $\nu = \omega/2\pi$. The second term dominates at a long distance, which is derived from the finite propagation time of the change of the electromagnetic field induced by the motion of charged particles. The electric field of radiation at a long distance is given by:

$$\vec{E_e} = -\nabla\Phi_e - \frac{\partial \vec{A}}{\partial t}$$

$$E_{xe} = \frac{q_e r_0 \omega}{4\pi\varepsilon}\left[\frac{(y^2 + z^2 - x^2)\sin\left[\omega\left(t - \frac{|\vec{r}|}{c_m}\right)\right] + 2xy\cos\left[\omega\left(t - \frac{|\vec{r}|}{c_m}\right)\right]}{|r|^4 c_m} + \frac{\omega\cos\left[\omega\left(t - \frac{|\vec{r}|}{c_m}\right)\right]}{|\vec{r}| c_m^2}\right]$$

$$E_{ey} = \frac{q_e r_0 \omega}{4\pi\varepsilon}\left[\frac{-2xy\sin\left[\omega\left(t - \frac{|\vec{r}|}{c_m}\right)\right] + (x^2 + z^2 - y^2)\cos\left[\omega\left(t - \frac{|\vec{r}|}{c_m}\right)\right]}{|r|^4 c_m} + \frac{\omega\sin\left[\omega\left(t - \frac{|\vec{r}|}{c_m}\right)\right]}{|\vec{r}| c_m^2}\right]$$

$$E_{ez} = \frac{q_e r_0 \omega}{4\pi\varepsilon}\left[\frac{-xz\sin\left[\omega\left(t - \frac{|\vec{r}|}{c_m}\right)\right] + yz\cos\left[\omega\left(t - \frac{|\vec{r}|}{c_m}\right)\right]}{|r|^4 c_m}\right]. \quad (3.55)$$

For a long distance, $\left|\vec{E_e}\right|$ and $\left|\vec{B_m}\right|$ are proportional to $r_0\omega^2/|\vec{r}|$. The radiation in the direction parallel to the motion of the charged particle gives the deceleration force because the radiation has momentum in the propagation direction. The expansion of the radiation energy is compensated by the loss of the kinetic energy of the charged particle. The radiation induced by the circular motion is called cyclotron radiation. With a given velocity $v_0 = r_0\omega$, the deceleration force by the radiation is inversely proportional to r_0^2. Therefore, a synchrotron accelerator with a large radius is required to attain a velocity close to the speed of light (Chapter 5.4).

3.4 Thermodynamics and statistical mechanics

What is thermal energy? From a microscopic perspective, the constituent atoms or molecules of an object have kinetic and potential energies. However, at the macroscopic scale, the kinetic energy of the object is given by that of the motion of the center of mass (Chapter 3.2). The relative motion between the constituent atoms or molecules provides an additional energy term called thermal energy. For example, we consider the kinetic energy of gaseous molecules with a mass of m. The total energy is given by

$$E_{tot} = \frac{m}{2}\sum_{i=1}^{N_g} v_i^2 \quad N_g\text{: number of molecules} \tag{3.56}$$

Considering the average velocity v_{ave} and $\delta v_i = v_i - v_{ave}$, (3.56) is rewritten as

$$E_{tot} = E_K + E_T \quad \left(\text{note } \sum_{i=1}^{N_g} v_{ave}(\delta v_i) = 0\right)$$

$$E_K = \frac{N_g m}{2} v_{ave}^2, \; E_T = \frac{m}{2}\sum_{i=1}^{N_g} (\delta v_i)^2 \tag{3.57}$$

is obtained. E_K is macroscopic kinetic energy, and E_T is thermal energy. For an observer moving with a certain velocity, E_K changes, but E_T does not change.

Macroscopic kinetic energy tends to transform into thermal energy. When an object slides on a floor, it is decelerated by friction, and the kinetic energy is transformed into thermal energy, which expands into the entire floor. Why does the matter on the floor not start to move, absorbing the thermal energy from the floor? This is because the probability of transforming into thermal energy is much higher than that for the inverse direction. Probability is the foundation of thermal mechanics and statistical mechanics.

Assuming the independence between different events, the probabilities are proportional to the number of states. We consider energies inside and outside the system E_{in} and E_{out}, where $E_{in} + E_{out} = E_{tot}$ is constant. The number of states inside and outside of the system are denoted by Ω_{in} and Ω_{out}, respectively. The possibility of energy distribution is proportional to

$$\Omega_{tot} = \Omega_{in}(E_{in})\Omega_{out}(E_{out}). \tag{3.58}$$

Entropy as a parameter of randomness is defined by $S_E = k_B \ln \Omega$, where k_B is the Boltzmann constant. Using entropy, (3.58) is rewritten as

$$S_{E-tot} = S_{E-in} + S_{E-out}. \tag{3.59}$$

The fundamental principle of statistical mechanics is the "increase of entropy":

$$\frac{dS_{E-tot}}{dt} = \frac{dS_{E-in}}{dt} + \frac{dS_{E-out}}{dt} \geq 0. \tag{3.60}$$

The thermal equilibrium state is attained when the total entropy is maximum. Equation (3.60) is rewritten as

$$\frac{dE_{in}}{dt}\left[\frac{dS_{E-in}}{dE_{in}}+\frac{dS_{E-out}}{dE_{in}}\right]=\frac{dE_{in}}{dt}\left[\frac{dS_{E-in}}{dE_{in}}-\frac{dS_{E-out}}{dE_{out}}\right]\geq 0. \tag{3.61}$$

Defining the thermodynamic temperature T by

$$\frac{1}{T}=\frac{dS_E}{dE}, \tag{3.62}$$

and (3.61) is rewritten as

$$\frac{dE_{in}}{dt}\left[\frac{1}{T_{in}}-\frac{1}{T_{out}}\right]\geq 0 \tag{3.63}$$

which indicates that E_{in} increases (decreases) when $T_{in} < T_{out}$ ($T_{in} > T_{out}$). With the thermal equilibrium state, $T_{in} = T_{out} (= T)$. The thermodynamic temperature was defined as a parameter that is uniform with thermal equilibrium.

Here we consider the distribution of each energy state, which is proportional to Ω_{tot} shown in (3.63). With the thermal equilibrium state, the following relation is derived.

$$\frac{dS_{E-out}}{dE_{in}}=-\frac{dS_{E-out}}{dE_{out}}=-\frac{1}{T}$$

$$S_{E-out}=k_B\ln\left(\Omega_{out}\right)=-\frac{E_{in}}{T}+const$$

$$\Omega_{out}\propto\exp\left(-\frac{E_{in}}{k_BT}\right)$$

$$\Omega_{tot}\propto\Omega_{in}(E_{in})\exp\left(-\frac{E_{in}}{k_BT}\right). \tag{3.64}$$

Here we consider the average of energy given by

$$E_{ave}=\int ER_{prob}(E)dx$$

$$R_{prob}(E)dx=\frac{\Omega(E)\exp\left(-\frac{E}{k_BT}\right)dx}{Z}$$

$$Z=\int\Omega(E)\exp\left(-\frac{E}{k_BT}\right)dx \quad x\text{: parametertogive energy.} \tag{3.65}$$

$R_{pob}(E)dx$ denotes the probability of having the energy of E. Z is a parameter to make $\int R_{pob}(E)dx = 1$. Seeing

$$E\exp\left(-\frac{E}{k_BT}\right) = -\frac{d}{d\left(\frac{1}{k_BT}\right)}\exp\left(-\frac{E}{k_BT}\right),$$

$$\int \Omega(E)E\exp\left(-\frac{E}{k_BT}\right)dx = -\frac{dZ}{d\left(\frac{1}{k_BT}\right)} \tag{3.66}$$

the average energy is given by

$$E_{ave} = -\frac{1}{Z}\frac{dZ}{d\left(\frac{1}{k_BT}\right)} = -\frac{d\ln(Z)}{d\left(\frac{1}{k_BT}\right)}. \tag{3.67}$$

The average one-dimensional kinetic energy of an object with the mass of m is given by

$$Z = \int \exp\left(-\frac{mv^2}{2k_BT}\right)dv = \sqrt{\frac{2\pi k_BT}{m}} \tag{3.68}$$

and the average of the one-dimensional kinetic energy is

$$K_{ave} = \frac{k_BT}{2}. \tag{3.69}$$

The average of the velocity in one direction is given by:

$$|v| - \sqrt{\frac{k_BT}{m}}. \tag{3.70}$$

The launching height with this initial velocity is given by (g: gravitational acceleration)

$$h_{launch} = \frac{v^2}{2g} = \frac{k_BT}{2mg}. \tag{3.71}$$

The launch height is inversely proportional to the mass. The mean velocity of gaseous molecular gas is the order of several hundred m/s, and the launched height is on the order of 4000 m. With $m = 10^{-3}$ kg, the launched height is below 10^{-18} m (four orders smaller than nuclei). However, we can consider that matter can consist of several mass components. Without binding, each mass component has kinetic energy given by (3.69) and can move at a higher velocity. Why does the velocity become slower owing to binding? The average velocity of the center of mass v_c of two particles with masses m_1 (velocity v_1) and m_2 (velocity v_2) is given by

$$v_c = \frac{m_1v_1 + m_2v_2}{m_1 + m_2}$$

$$\langle v_c^2 \rangle_{ave} = \frac{m_1^2 \langle v_1^2 \rangle_{ave} + 2m_1 m_2 \langle v_1 v_2 \rangle_{ave} + m_2^2 \langle v_2^2 \rangle_{ave}}{(m_1 + m_2)^2} = \frac{k_B T}{m_1 + m_2}.$$

$$\left(\text{using } \langle v_{1,2}^2 \rangle_{ave} = \frac{k_B T}{m_{1,2}}, \ \langle v_1 v_2 \rangle_{ave} = 0 \right) \tag{3.72}$$

The ideal gas law (Boyle and Charles's law, shown in Chapter 1.6) is given by

$$PV = N_g k_B T \tag{3.73}$$

where P is the pressure, V is the volume, and N_m is the number of gaseous molecules. Equation (3.73) is derived statically mechanically as follows. The gas pressure is derived from the impact force at the collision of gas molecules on the wall. Considering the mass and velocity of the gas molecules m and v, the change of the momentum (impact force) with a single collision is given by

$$p = mv \rightarrow -mv$$

$$\Delta p = 2mv. \tag{3.74}$$

On the unit area of the wall, the collision number of times with the gas molecules per unit time is given by

$$\Gamma_{col-wall} = \frac{N_g v}{2V} \text{ (half of molecules have motion to the wall).} \tag{3.75}$$

Then the pressure is given by

$$P = \Delta p \times \Gamma_{col-wall} = mv^2 \frac{N_g}{V}. \tag{3.76}$$

Seeing (3.69), the average of mv^2 is $k_B T$, and (3.73) is derived.

3.5 Mysteries of light and the dawn of modern physics

The elucidation of the identity of light caused the appearance of modern physics: the theory of relativity and quantum mechanics. As shown below, the speed of light in a vacuum is constant. And light has a dual nature, such as waves and particles. There was a mystery as to why light has such special characteristics. But the theory of relativity indicates that all particles without mass must move at the speed of light. Quantum mechanics indicates that all particles have a wave nature.

3.5.1 Constancy of the speed of light

Chapter 3.3 indicates the identity of light as the electromagnetic wave, and the propagation speed was obtained as $c_m = 1/\sqrt{\varepsilon\mu}$. The observed velocity depends on the motion of observers. We assume that the velocity of $\vec{V}$ is observed by A. When

B is moving with the velocity of $\vec{v}$ against A, B would observe the velocity of $\vec{V} - \vec{v}$. For the propagation in a medium, c_m is the velocity against the medium. For propagation in a vacuum, what is the medium that serves as the standard for the speed of light? The values of ε and μ in a vacuum space (ε_0 and μ_0) are universal constants, and the speed of light in a vacuum $c = 1/\sqrt{\varepsilon_0\mu_0}$ derived from Maxwell's equation does not depend on the motion of the observer. To solve this discrepancy, the speed of light in a vacuum was defined as the speed relative to the medium called ether. If we move against the ether, the speed of light is expected to shift. However, this effect was estimated to be too small to be detected because we only experience velocities much slower than the speed of light. The speed of light was measured from the round-trip propagation time. The influence of the motion with the speed of v on the round-trip propagation time for the distance of L is given by

$$T_{prop} = \frac{L}{c-v} + \frac{L}{c+v} \approx \frac{2L}{c}\left[1 + \left(\frac{v}{c}\right)^2\right]. \tag{3.77}$$

For example, the speed of the Earth revolving around the Sun is on the order of 10^{-4}; then this effect is on the order of 10^{-8} and could not be detected with the measurement uncertainty of the speed of light in the nineteenth century.

Michelson and Morley expected that this effect could be detected by comparing the round propagation times in the parallel and perpendicular directions to the Earth's orbital motion (v_{orb}) as shown in Figure 3.4 [5]. The round propagation times in the parallel and perpendicular directions are given by $\tau_{\parallel} = 2L_{\parallel}c/\left(c^2 - v_{orb}^2\right)$ and $\tau_{\perp} = 2L_{\perp}/\sqrt{c^2 - v_{orb}^2}$, respectively ($L_{\parallel,\perp}$: distance to the mirror in the parallel and perpendicular directions). With the orbital motion of the

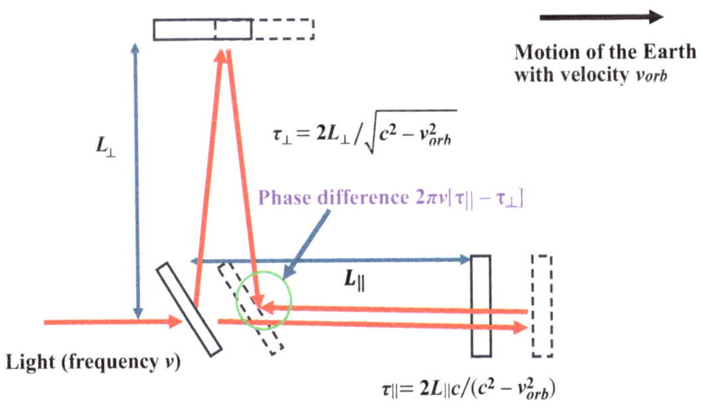

Figure 3.4 The apparatus for Michelson and Morley's experiment to observe the influence of the orbital motion of the Earth on the speed of light from the change of the phase shift between the lights' parallel and perpendicular directions

Earth (change of direction of motion), the variation in $\Delta\tau = (\tau_{\parallel} - \tau_{\perp})$ was expected to be observed by the variation in the interference signal between the lights round propagated in both directions. The interference signal indicates a phase difference between two optical paths, $\Delta\varphi_{ph} = 2\pi\nu(\Delta\tau)$, where ν is the frequency of the light (with this experiment, a Na lamp with the wavelength of 589 nm). With Michelson and Morley's experiment, $L_{\parallel} = L_{\perp} = 11$ m. The observation of the interference signal is advantageous to observe the slight variation of $\Delta\tau$, because it focuses after the decimal point of $\Delta\varphi_{ph}/2\pi = \nu(\Delta\tau)$. The change of speed of light with the ratio of 10^{-8} was expected to be detected with the phase change of 0.37 radian. The variation in the speed of light according to the Earth's orbital motion (speed of $10^{-4}c$) was expected to be detected with this system. However, the speed of light was measured to be constant within a fractional uncertainty of 10^{-8}. Using a laser source, the constancy of the speed of light has been confirmed with an uncertainty of 10^{-15} [6]. The constancy of the speed of light remained a mystery until the establishment of the theory of relativity. The speed of light was defined to be 299 792 458 m/s in 1983.

3.5.2 The particle nature of light

The identity of light was clarified to be the electromagnetic wave. However, some phenomena could not be explained by considering light only as a wave. When a material is heated, emission of light is observed: this irradiation is called "black body radiation." White thermoelectric balls are light sources using blackbody radiation. The following characteristics were experimentally obtained with the blackbody radiation (shown in Figure 3.5) [7].

1. The intensity is maximum at a certain frequency, which is proportional to the thermodynamic temperature T (Chapter 3.4), and the total intensity is proportional to T^4. The blackbody radiation at room temperature is not visible because it is distributed mostly in the infrared region (visible using an IR viewer).

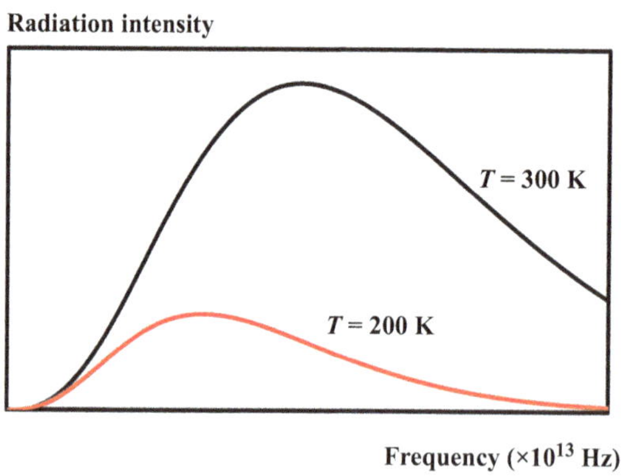

Figure 3.5 Intensity distribution of blackbody radiation

2. The energy density at the low frequency region is given by $\frac{8\pi\nu^2}{c^3}k_BT$, where ν is the frequency, and k_B is the Boltzmann constant (1.38×10^{-23} J/K). This characteristic is explained by the pure wave nature of radiation.
3. With the high-frequency region, the distribution is proportional to $\nu^3\exp\left[-\frac{h\nu}{k_BT}\right]$. Here, h is the Planck constant (6.6×10^{-34} J/Hz). This characteristic cannot be explained considering light only as a wave.

These results indicate that the properties of light cannot be described as a pure wave nature with high frequency. Planck derived a general formula of the spectrum distribution as follows, based on the assumption that the energy can only assume $n_\alpha h\nu$ with the probability proportional to $\exp\left(-\frac{n_\alpha h\nu}{k_BT}\right)$ (see Chapter 3.4), where $n_\alpha(\geq 0)$ is an integer.

$$P_{BBR}d\nu = \frac{8\pi h\nu^3}{c^3}\frac{1}{\exp\left[\frac{h\nu}{k_BT}\right] - 1}d\nu, \tag{3.78}$$

which is in good agreement with the experimental result in the whole frequency area. Taking $x = h\nu/k_BT$, (3.78) is rewritten as

$$\int P_{BBR}d\nu = \frac{8\pi x^3}{h^3c^3}(k_BT)^4\int\frac{x^3}{\exp[x] - 1}dx \tag{3.79}$$

and the total intensity is proportional to T^4 as shown in result (1). P_{BBR} takes a maximum with a certain value of x, which is proportional to T. With low frequency, the result (2) is derived from an approximation of $\exp(h\nu/k_BT) - 1 \approx h\nu/k_BT$. With the high frequency, the result (3) is derived from the approximation $\exp(h\nu/k_BT) - 1 \approx \exp(h\nu/k_BT)$. However, the validity of Planck's assumption was not confirmed at that time.

Another mysterious result was observed with the photoelectronic effect: the emission of electrons from a solid surface induced by the irradiation of light, as shown in Figure 3.6 [8]. The experimental results show that the energy of the emitted electron

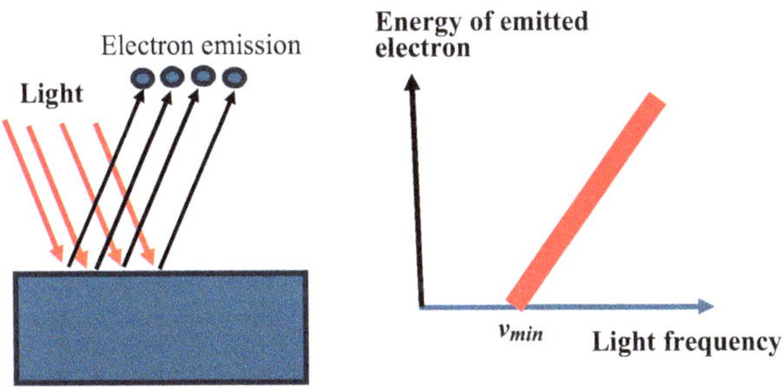

Figure 3.6 Photoelectronic effect, which indicates the emission of electrons by irradiation of light. This emission is not caused when the frequency of light ν is below ν_{min}. The energy of the emitted electron is proportional to $\nu - \nu_{\text{min}}$.

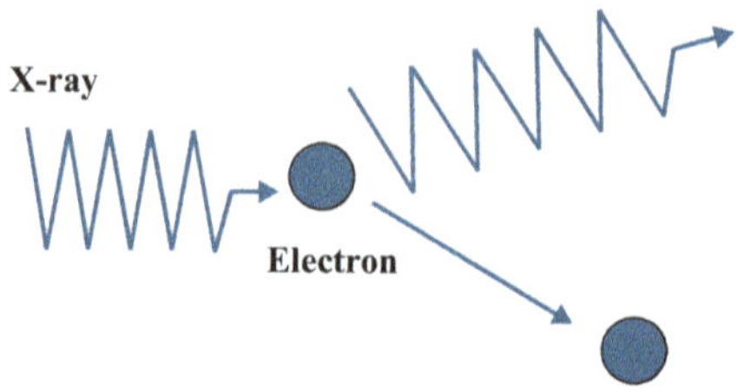

Figure 3.7 Compton scattering indicating that X-rays can scatter electrons, changing wavelength and conserving the total momentum of the photon and electron

is independent of the intensity of the light, although the number of emitted electrons is proportional to the intensity. Moreover, emissions do not occur when the frequency of light ν is lower than a minimum threshold value ν_{min}, and the energy of the emitted electron is proportional to the $\nu - \nu_{\min}$ (Figure 3.6). Einstein proposed a new concept of wave-particle duality; light has the nature of both waves and particles [9]. The energy of each particle (called a photon) is $E = h\nu$, and the momentum is $\vec{p} = h\vec{k}$ ($|\vec{k}| = 1/\lambda$), where $\vec{k}$ is the wavenumber vector and λ is the wavelength.

This theory was confirmed by Compton scattering, indicating that X-rays can cause the acceleration of electrons. The wavelength of the X-ray is increased so that the total momentum of the photon and electron is conserved, as shown in Figure 3.7 [10].

Planck's assumption was also explained by the particle-wave duality of light: light energy is the product of the photon number n_α and the energy of each photon $h\nu$. This duality was interpreted as a special characteristic of light until the concept of matter waves was proposed.

3.6 Special relativity

Several hypotheses were given to interpret the result of Michelson and Morley, for example, the compression of ether by the motion, etc. In 1905, Einstein established the theory of special relativity based on the constant speed of light [11]. When an observer is moving in the x-direction with velocity v, the coordinate of the position and time (x,y,z, ct) is observed as (x',y',z',ct'), which are given by the relations (Lorentz transformation)

$$\begin{pmatrix} x' \\ y' \\ z' \\ ct' \end{pmatrix} = \begin{pmatrix} \frac{1}{\sqrt{1-(v/c)^2}} & 0 & 0 & \frac{-(v/c)}{\sqrt{1-(v/c)^2}} \\ 0 & 1 & 0 & 0 \\ 0 & 0 & 1 & 0 \\ \frac{-(v/c)}{\sqrt{1-(v/c)^2}} & 0 & 0 & \frac{1}{\sqrt{1-(v/c)^2}} \end{pmatrix} \begin{pmatrix} x \\ y \\ z \\ ct \end{pmatrix} \tag{3.80}$$

Equation (3.80) is derived from the equivalency between $(ct)^2 = x^2 + y^2 + z^2$ and $(ct')^2 = x'^2 + y'^2 + z'^2$. The length in a moving coordinate is given by

$$x_1' - x_2' = \frac{1}{\sqrt{1 - \left(\frac{v}{c}\right)^2}}[(x_1 - vt_1) - (x_2 - vt_2)]$$

when $t_1' = t_2'$, $t_1 - t_2 = \frac{v}{c^2}(x_1 - x_2)$

$$x_1' - x_2' = \sqrt{1 - \left(\frac{v}{c}\right)^2}(x_1 - x_2) \tag{3.81}$$

The simultaneous phenomena at distant positions $(t_1 = t_2,\ x_1 \neq x_2)$ are not simultaneous in a moving coordinate frame $(t_1' \neq t_2')$. A hypothesis that the length in the direction parallel to the motion becomes shorter with a factor of $\sqrt{1 - (v/c)^2}$ was given to explain the result of Michelson and Morley's experiment (Chapter 3.5.1). Compression of ether by motion was considered the cause, but there was no explanation about this factor. From the theory of relativity, this result is derived from the constancy of the speed of light. The time progress $t_1 - t_2$ is transformed as follows

using $x_1' = x_2'$,

$$t_1' - t_2' = \sqrt{1 - \left(\frac{v}{c}\right)^2}(t_1 - t_2). \tag{3.82}$$

This slower time progress is interpreted by comparing the propagation time of light for the distance of L. The propagation time is L/c without any motion. However, if it moves with the velocity of v in the direction perpendicular to the light propagation, the light has a velocity component of v parallel to the motion, but the absolute value of the speed of light is constant. The propagation time is $L/\sqrt{c^2 - v^2}$, as shown in Figure 3.8.

The Lorenz transform was proposed by Lorenz. But Lorenz defined v as the velocity relative to the ether, while it is relative velocity with the theory of relativity.

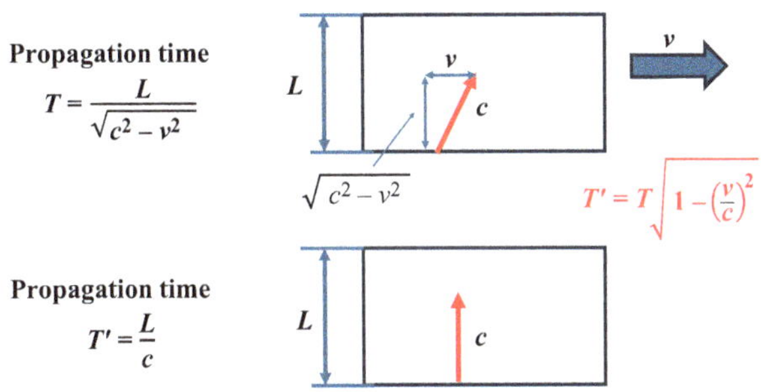

Figure 3.8 The comparison of light propagation time with and without the motion perpendicular to the propagation direction. With the motion, the time progress is slower.

Here, we consider the case that A and B are moving with a certain relative velocity. From the theory of relativity, A observes a shorter length and slower time progress with B, and B observes the same with A. But from Lorenz's theory, this relation is not valid.

The velocity component (V_x, V_y, V_z) in the moving coordinate system is given by:

$$V'_x = \frac{dx'}{dt'} = \frac{d\left[\frac{1}{\sqrt{1-\left(\frac{v}{c}\right)^2}}(x-vt)\right]}{dt}\frac{dt}{dt'}$$

taking $\frac{dx}{dt} = V_x$

$$\frac{dt'}{dt} = \frac{d}{dt}\left[\frac{1}{\sqrt{1-\left(\frac{v}{c}\right)^2}}\left(t-\frac{vx}{c^2}\right)\right] = \frac{1}{\sqrt{1-\left(\frac{v}{c}\right)^2}}\left(1-\frac{vV_x}{c^2}\right)$$

$$V'_x = \frac{V_x - v}{1-\frac{vV_x}{c^2}}$$

$$V'_y = \frac{dy'}{dt'} = \frac{dy}{dt}\frac{dt}{dt'} = \sqrt{1-\left(\frac{v}{c}\right)^2}\frac{V_y}{1-\frac{vV_x}{c^2}}$$

$$V'_z = \sqrt{1-\left(\frac{v}{c}\right)^2}\frac{V_z}{1-\frac{vV_x}{c^2}}. \tag{3.83}$$

Here, we consider the case $V_x = c\cos\theta$ and $V_y = c\sin\theta$. Then

$$V'_x = \frac{c\cos\theta - v}{1-\frac{v}{c}\cos\theta}, \quad V'_y = \sqrt{1-\left(\frac{v}{c}\right)^2}\frac{c\sin\theta}{1-\frac{v}{c}\cos\theta}$$

$$V'_x2 + V'_y2 = c^2 \tag{3.84}$$

and the absolute value of the speed of light is constant for any moving frame.

The conservation of momentum $\vec{p}$ and energy E $(\vec{p_1}+\vec{p_2}=\vec{p_3}+\vec{p_4},\ E_1+E_2=E_3+E_4)$ is fundamental to mechanics, and we assume that it holds in any coordinates. For classical mechanics, momentum is defined as $\vec{p}=m\vec{V}$ (m: mass, $\vec{V}$: velocity) as shown in Chapter 3.2.1. In another coordinate, the momentum should be given by $\vec{p'}=m\vec{V'}$ ($\vec{V'}$ is obtained from (3.83)), with which the conservation of momentum does not hold. We need another definition of momentum that converges to the momentum of classical mechanics for $|\vec{V}| \ll c$. We consider the four-dimensional momentum vectors $\vec{p^4} = (p_x, p_y, p_z, E/c)$. The equivalent relation between $(E/c)^2 = p_x^2 + p_y^2 + p_z^2$ and $(E'/c)^2 = p'^2_x + \mathrm{p}'^2_y + p'^2_z$

holds for photons (light), and the transform between different coordinates is given by

$$\begin{pmatrix} p'_x \\ p'_y \\ p'_z \\ E'/c \end{pmatrix} = \begin{pmatrix} \frac{1}{\sqrt{1-(v/c)^2}} & 0 & 0 & \frac{-(v/c)}{\sqrt{1-(v/c)^2}} \\ 0 & 1 & 0 & 0 \\ 0 & 0 & 1 & 0 \\ \frac{-(v/c)}{\sqrt{1-(v/c)^2}} & 0 & 0 & \frac{1}{\sqrt{1-(v/c)^2}} \end{pmatrix} \begin{pmatrix} p_x \\ p_y \\ p_z \\ E/c \end{pmatrix}. \quad (3.85)$$

Using this relation, $\vec{p_1^4} + \vec{p_2^4} = \vec{p_3^4} + \vec{p_4^4}$ and $\vec{p_{1'}^4} + \vec{p_{2'}^4} = \vec{p_{3'}^4} + \vec{p_{4'}^4}$ are equivalent, and the conservation of the total momentum and energy is guaranteed in any coordinate frame. We consider the transformation from $\vec{p^4} = (0, 0, 0, E/c)$,

$$p'_x = -\frac{-\frac{v}{c^2}E}{\sqrt{1-\left(\frac{v}{c}\right)^2}} = \frac{\frac{V'_x}{c^2}E}{\sqrt{1-\left(\frac{V'_x}{c}\right)^2}} \quad (V'_x = -v) \quad (3.86)$$

$$E' = \frac{E}{\sqrt{1-\left(\frac{V'_x}{c}\right)^2}}. \quad (3.87)$$

Assuming the rest energy (energy with $|\vec{p}| = 0$) of $E = mc^2$,

$$p'_x = \frac{mV'_x}{\sqrt{1-\left(\frac{V'_x}{c}\right)^2}}, \quad (3.88)$$

$$E' = \frac{mc^2}{\sqrt{1-\left(\frac{V'_x}{c}\right)^2}}. \quad (3.89)$$

For $V'_x \ll c$, (3.88) and (3.89) are approximate to:

$$p'_x = mV'_x$$
$$E' = mc^2 + \frac{m}{2}V'^2_x. \quad (3.90)$$

Equation (3.90) shows the momentum for classical mechanics, and the energy is the sum of the rest energy and kinetic energy in the case of classical mechanics. Considering nonzero $V_{x,y,z}$, four-dimensional momentum vector $(p_x, p_y, p_z, E/c)$ is defined by

$$p_{x,y,z} = \frac{mV_{x,y,z}}{\sqrt{1-\left(V_x^2 + V_y^2 + V_z^2\right)/c^2}} \quad (3.91)$$

$$E = \frac{mc^2}{\sqrt{1-\left(V_x^2+V_y^2+V_z^2\right)/c^2}}. \tag{3.92}$$

Equations (3.91) and (3.92) indicate that the motion equation in each direction is not independent of the theory of relativity. Therefore, the two-body motion (shown in Chapter 3.2.1) is not separated into the motion of the center of mass and the relative motion.

Equation (3.92) indicates that the energy of a body with nonzero mass becomes infinite when $v = c$. Therefore, a particle with nonzero mass cannot be accelerated to a velocity higher than the speed of light. On the other hand, massless particles can move only at the speed of light. Comparing (3.91) and (3.92),

$$E^2 = m^2c^4 + c^2\left(p_x^2+p_y^2+p_z^2\right), \tag{3.93}$$

is derived. Equation (3.93) indicates $E = c|\vec{p}|$ with $m = 0$, which was confirmed with photons as shown in Chapter 3.3.3. Equation (3.93) is valid for particles with and without mass.

With the electromagnetic field, the $\left(\overrightarrow{p_Q} \rightarrow \overrightarrow{p_Q} + q_e\overrightarrow{A_Q},\ E \rightarrow E - q_e\Phi_e,\ Q = x, y, z\right)$ transform is performed as shown in Chapter 3.2.2, where q_e is the electric charge, $A_{x,y,z}$ is the vector potential of the magnetic field, and Φ_e is the electric voltage. The magnetic field is given by $\overrightarrow{B_m} = \nabla \times \vec{A}$ and the electric field is given by $\overrightarrow{E_e} = -\nabla\Phi_e - \partial\vec{A}/\partial t$. The Lorentz transformation of the four-dimensional momentum vector should also be valid with the electromagnetic field, and the Lorentz transform of $\vec{A_4} = (A_x, A_y, A_z, -\Phi_e/c)$ is given as follows.

$$\begin{pmatrix} A'_x \\ A'_y \\ A'_z \\ -\Phi'_e/c \end{pmatrix} = \begin{pmatrix} \frac{1}{\sqrt{1-(v/c)^2}} & 0 & 0 & \frac{-(v/c)}{\sqrt{1-(v/c)^2}} \\ 0 & 1 & 0 & 0 \\ 0 & 0 & 1 & 0 \\ \frac{-(v/c)}{\sqrt{1-(v/c)^2}} & 0 & 0 & \frac{1}{\sqrt{1-(v/c)^2}} \end{pmatrix} \begin{pmatrix} A_x \\ A_y \\ A_z \\ -\Phi_e/c \end{pmatrix} \tag{3.94}$$

Assuming $A_{x,y,z} = 0$ and $\Phi_e \neq 0$ in a frame,

$$\left(A'_x, A'_y, A'_z, -\frac{\Phi'_{el}}{c}\right) = \frac{1}{\sqrt{1-(v/c)^2}}\left(\frac{v\Phi_{el}}{c^2}, 0, 0, -\frac{\Phi_e}{c}\right) \tag{3.95}$$

is derived in another frame moving in the x-direction with the velocity of v. The magnetic field is given by

$$B'_{my} = \frac{\partial A'_x}{\partial z'} = \frac{\partial A'_x}{\partial z} = -\frac{1}{\sqrt{1-(v/c)^2}}\frac{v}{c^2}E_{ez}$$

$$\overrightarrow{B'_m} = \frac{1}{\sqrt{1-(v/c)^2}}\left(0, -\frac{v}{c^2}E_{ez}, \frac{v}{c^2}E_{ey}\right). \tag{3.96}$$

The electric current is not zero in a moving frame, and the magnetic field is also nonzero. Assuming the linear electric charge of σ_{el} in a unit length, the electric current is given by $I_{el} = \sigma_{el}v$. The electric field is given by

$$E_{ey,z} = \frac{\sigma_{el}}{2\pi\varepsilon_0(y^2+z^2)}y, z. \tag{3.97}$$

Substituting (3.97) into (3.96),

$$\overrightarrow{B'_m} = \frac{1}{\sqrt{1-(v/c)^2}}\left(0, -\frac{\mu_0 I_{el} z}{2\pi(y^2+z^2)}, \frac{\mu_0 I_{el} y}{2\pi(y^2+z^2)}\right) \tag{3.98}$$

is obtained, which corresponds to Ampere's circuital law with $v \ll c$.

The relation shown above is consistent, considering the Lorenz transform of four-dimensional electric current $\overrightarrow{j_4} = (j_x, j_y, j_z, c\rho_e)$ as follows:

$$\begin{pmatrix} j'_x \\ j'_y \\ j'_z \\ c\rho'_e \end{pmatrix} = \begin{pmatrix} \frac{1}{\sqrt{1-(v/c)^2}} & 0 & 0 & \frac{-(v/c)}{\sqrt{1-(v/c)^2}} \\ 0 & 1 & 0 & 0 \\ 0 & 0 & 1 & 0 \\ \frac{-(v/c)}{\sqrt{1-(v/c)^2}} & 0 & 0 & \frac{1}{\sqrt{1-(v/c)^2}} \end{pmatrix} \begin{pmatrix} j_x \\ j_y \\ j_z \\ c\rho_e \end{pmatrix}, \tag{3.99}$$

where $\vec{j}$ is the electric current density vector, and ρ_e is the charge density. The transformation from $\overrightarrow{j_4} = (0, 0, 0, c\rho_{e0})$ is given by $\overrightarrow{j_4} = (\rho_{e0}v, 0, 0, c\rho_{e0})/\sqrt{1-(v/c)^2}$ as shown in Figure 3.9. The density is increased by the reduced size in the direction of motion. Using (3.99), (3.3.12) is valid with any coordinate.

The Lorentz transform converges to the Galilei transform $(x' = x - vt)$ with $v \ll c$. However, the Galilei transform was valid only for $\vec{r_4}$. By the introduction of the rest energy mc^2, Lorentz transform became applicable also for $\overrightarrow{p_4}, \overrightarrow{A_4}$, and $\overrightarrow{j_4}$.

There is a paradox with the total torque giving the forces F at two points A and B, as shown in Figure 3.10. The lengths A-O and B-O are assumed to be equal (L_p) for the observer with the zero relative velocity to this object. Then the total torque is zero, and no rotation is caused. The force is the time differential of momentum $(\text{inthisframe}, F = dp_{A,B}/dt)$. For an observer with the relative velocity of v in the B-O direction, the forces at the A and B points are given by $dp'_A/dt' = F$ and $dp'_B/dt' = F\sqrt{1-\left(\frac{v}{c}\right)^2}$, respectively, according to (3.80) and (3.85). The A-O length is L_p, while the B-O length is $L_p\sqrt{1-(v/c)^2}$. There is a total torque of

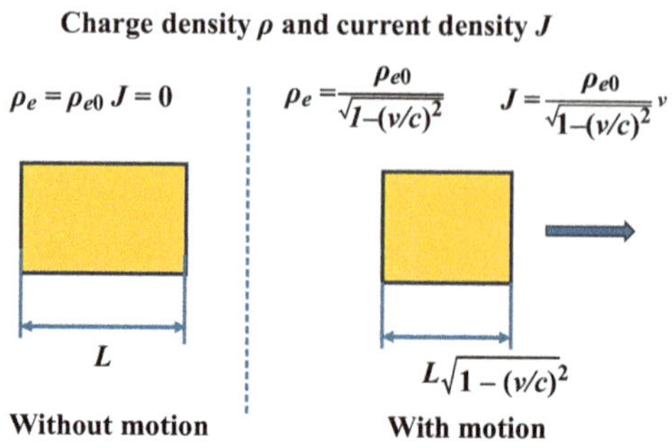

Figure 3.9 *The Lorenz transform of electric charge without motion (electric current is zero). Note that the reduction of the size in the motion direction increases the charge density and current density.*

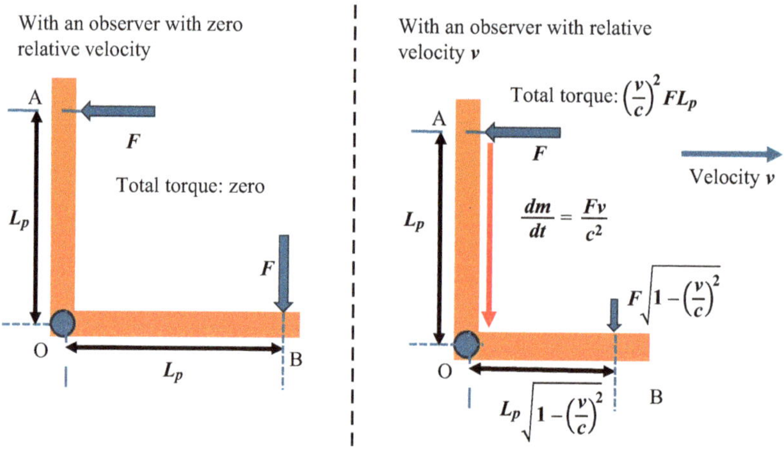

Figure 3.10 *A paradox with the balance of total torque. For the observer with zero relative velocity, the total torque is zero, and the object does not rotate. However, for an observer with nonzero relative velocity in the B-O direction, the total torque is nonzero. There is a question of why the object does not rotate. The solution is given in the text.*

$(v/c)^2 FL_p$. Then the object should rotate? This discrepancy is solved by considering that the torque gives the change of the angular momentum, which is the product of the inertial moment and the angular velocity. The angular moment can change without a change of angular velocity when there is a change of inertial moment, as shown below. The force at A gives a change of energy with the rate of $dE/dt = Fv$. The change of energy gives a change of effective mass by $dm/dt = Fv/c^2$. The energy flows by $A \rightarrow$ O, and it leaks at O. The change of the inertial moment per unit time is given by $dI_{in}/dt = (dm/dt)L_p^2$, and the angular velocity is given by v/L_p. The change of the angular momentum per unit time is given by $(dm/dt)L_p^2(v/L_p) = (v/c)^2FL_p$. There is an effective change of mass at point O in the inverse direction; therefore, there is no change of total mass. The change of effective mass at point O does not have any effect on the change of the inertial moment. Therefore, there is no discrepancy between the change of the angular momentum and the total torque with any moving frame.

3.7 General relativity

We consider the case that A and B are moving with a high relative velocity. With the theory of special relativity, the relativistic effect that A observes with B is the same as that which B observes with A. For example, A observes slower time progress with B, while B also observes slower time progress with A. When they meet again, which one is younger? This question is meaningless because they cannot meet again without any acceleration, while the theory of special relativity is valid only in the inertial frame of reference. When only A gets acceleration, A feels the inertial force, but B does not. Then the observation of A by B and the observation of B by A are not equal.

The theory of general relativity was developed to discuss the mechanics of acceleration, based on the equality of inertial force and gravitational force [11]. When we are in an accelerated system, we see that the optical path is bent (Figure 3.11). Einstein expected that the bending of the optical path would also be caused by gravity. Then the position of stars should be observed at a shifted position when the optical path is bent by passing near a massive star. This expectation was confirmed in 1919 by Eddington when observing a distant star during a solar

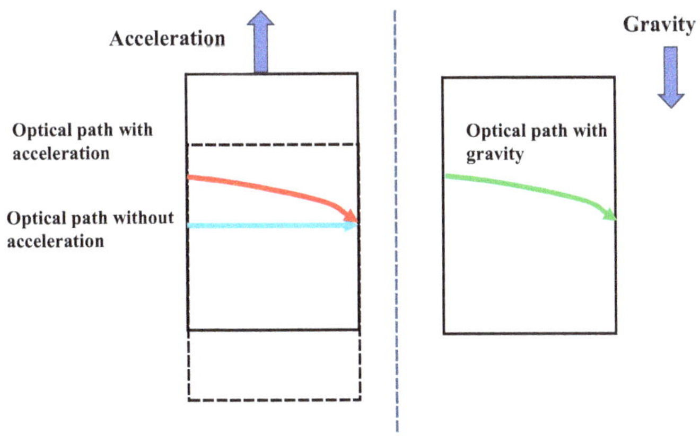

Figure 3.11 Bending of an optical path in an accelerated frame of reference. Einstein considered the same phenomenon also with gravity.

eclipse [12]. There is also a phenomenon called "gravitational lens," with which the light from distant stars is focused by massive materials on the path to the Earth. The mass distribution can be measured using the bend of the optical path, also when the mass is dominated by dark matter, which is not visible (Chapter 6.6).

With Newtonian mechanics, gravity is an attractive force between matter having mass. For matters with initial velocity v in the horizontal direction, the velocity components in the horizontal and vertical directions are v and gt (g: gravitational acceleration), respectively. This concept is not valid for the theory of general relativity, because the bent path is also caused by light, which has no mass. The orbit of matter given with Newtonian mechanics is valid only when $v \ll c$. The speed of light is constant also with the gravitational potential, and the horizontal and vertical velocity components of light are $\sqrt{c^2 - (gt)^2}$ and gt, respectively (also for particles moving with velocity close to c). Comparing the propagation time in the horizontal direction with and without the gravitational potential, the progress of time becomes slower by the gravitational potential with a factor of $\sqrt{1 - (gt/c)^2}$ (gravitational redshift). It is not simple to obtain the formula for the redshift observed with the general formula of gravitational potential. Here we consider the change of progress of time between the change of altitude $h_G \rightarrow h'_G (h'_G > h_G)$. Using $h'_G - h_G = gt^2/2$,

$$\Delta t(h'_G) = \sqrt{1 + \frac{2g(h'_G - h_G)}{c^2}} \Delta t(h_G) \approx \left[1 + \frac{g(h'_G - h_G)}{c^2}\right] \Delta t(h_G) \qquad (3.100)$$

On the Earth, the progress of time becomes faster with a change of altitude of 1 m with the ratio of 1.1×10^{-16}.

When $gt > c$, the horizontal velocity component of the light is imaginary. Taking the size of the gravitational potential area D_{BL}, the light cannot propagate in the horizontal direction at $t > D_{BL}/c$ when

$$g\frac{D_{BL}}{c^2} > 1. \qquad (3.101)$$

When (3.101) is satisfied, any matter with the speed of light cannot escape from the gravitational potential area, which is called a "black hole." Equation (3.101) indicates that matter with any mass of m can be a black hole when it is compressed to a radius below $R_{SW} (< D_{BL}/2)$ satisfying

$$g = \frac{Gm}{R_{SW}^2}$$

$$R_{SW} = \frac{2Gm}{c^2}. \qquad (3.102)$$

R_{SW} obtained by (3.102) is called "Schwarzshild radius." With solar mass, the Schwarzshild radius is 3 km. The existence of a black hole was confirmed from the revolution motion of stars around a point where no radiation is observed, as shown

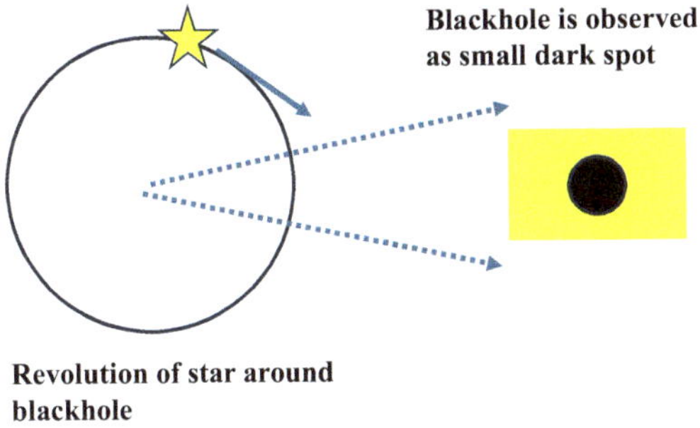

Figure 3.12 Discovery of black holes from the revolution of stars around the position where no radiation is observed. The shadow of a black hole is observed with ultra-high spatial resolution.

in Figure 3.12. The sights of black holes are dark shadows, which are difficult to observe with a single telescope because of their small size. Increasing the spatial resolution by taking the interference of signals from telescopes in Arizona, Chile (two), Mexico, Hawaii (two), Spain, and the South Pole, the shadow of the black hole (mass of 6.5 billion times the solar mass) in M87 (55 million light-years distant) was observed [13]. The discovery of black holes confirmed the validity of the theory of general relativity.

Black holes with a mass of 20–100 times the solar mass are called "stellar black holes," which are constructed by the collapse of the cores of stars. On the other hand, black holes with a mass of 10^5–10^9 times the solar mass (called "supermassive black holes") have been discovered at the center of large galaxies. It is a mystery how these supermassive black holes came to be. It is reasonable to consider that stellar black holes can grow by absorbing surrounding objects or merging with other black holes. This procedure is expected to take several billion years. However, supermassive black holes were discovered in very distant galaxies, indicating that supermassive black holes existed within 700 million years after the birth of the universe (Chapter 6.4). Supermassive black holes were constructed within a short period with an unknown mechanism. And it is also a mystery that black holes with a mass in the range of 10^2–10^5 times the solar mass have never been discovered [14]. If stellar black holes grow to supermassive black hole, there must be a procedure for being the intermediate mass. There might be a limit on the mass that can be obtained by growing stellar black holes, which is not clarified yet.

With the theory of general relativity, gravity is not an attractive force but the distortion of space. Matters move uniformly along the bent axis. The gravitational force is not exactly inversely proportional to the square of distance. The perihelion shift (rotation of the long axis of the elliptical orbit) of Mercury was estimated accurately by the theory of general relativity [15], while there was a discrepancy of 8% between the observation and the estimation by Newtonian mechanics.

When there is a temporal change of the gravitational potential distribution, the change of the distortion of the space propagates as a wave with the speed of light, which is called a "gravitational wave," just like the motion of a charged particle gives an irradiation of an electromagnetic wave (Chapter 3.3.4). The gravitational wave propagating in the z-direction gives the change of length of matter in the x- and y-directions by

$$L_x(t,z) = L_{x0}(z)\left\{1 + \delta_G \sin\left[2\pi\nu_R\left(t - \frac{z}{c}\right)\right]\right\}$$
$$L_y(t,z) = L_{y0}(z)\left\{1 - \delta_G \sin\left[2\pi\nu_R\left(t - \frac{z}{c}\right)\right]\right\}, \tag{3.103}$$

where ν_R is the frequency. When the gravitational wave is induced by a revolution motion of binary stars, ν_R indicates the frequency of the revolution motion. The observation of change of length by the gravitational wave has been very difficult, because δ_G is below 10^{-21} also by the merge of black holes (it is the ratio between the size of an atom and the distance between the Sun and the Earth).

3.8 Quantum mechanics

3.8.1 Early quantum theory

Chapter 1.7 indicates the procedure to recognize the atomic structure with nuclei and electrons revolving around them. However, the stable revolution motion of an electron could not be explained with Newtonian mechanics and electromagnetism. A novel physics was required to describe the phenomena at the micro level.

There was another mystery that only the frequency ν_{rad} satisfying (Figure 3.13)

$$\frac{\nu_{rad}}{c} = Ry\left[\frac{1}{n_1^2} - \frac{1}{n_2^2}\right] \tag{3.104}$$

$n1, 2$: integer Ry: Rydberg constant

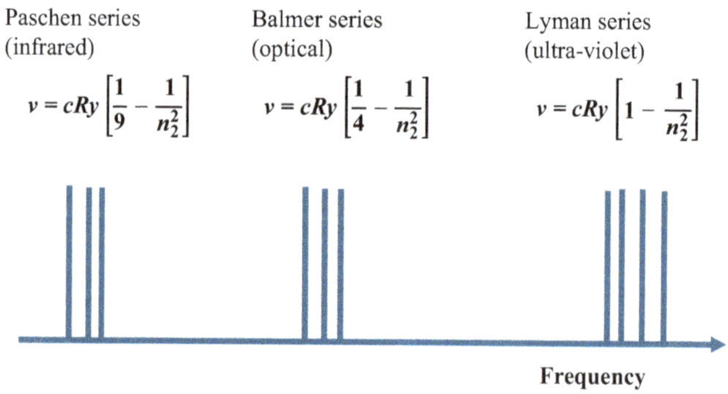

Figure 3.13 The observed frequency of emission from hydrogen atoms

were emitted from hydrogen atoms [16,17]. The Lyman series (ultraviolet), Balmer series (optical), and Paschen series (infrared) indicate the series with $n_1 = 1, 2$, and 3, respectively.

In 1913, Bohr established the "Old Quantum Mechanics" from the assumption that particles bounded in a limited area of q must satisfy [18]

$$\oint p_q dq = nh p_q\text{: momentum in the } q - \text{direction } n\text{: integer.} \tag{3.105}$$

With this requirement, the radius of the electron orbit with velocity v in a hydrogen atom is limited to

$$2\pi\mu_e vr = nh(\text{momentum is } \mu_e v)$$

$$\frac{e^2}{4\pi\varepsilon_0 r^2} = \frac{\mu_e v^2}{r}$$

(balance between Coulomb force and centrifugal force shown in (3.15))

$$r = a_B n^2$$

$$a_B = \frac{\varepsilon_0 h^2}{\pi\mu_e e^2}$$

$$\mu_e\left(= \frac{m_e m_p}{m_e + m_p}\right)\text{: reduced mass between electron } (\text{mass } m_e) \text{ and proton } (\text{mass } m_p) \tag{3.106}$$

Thc possible electron energy is given by

$$E_{en} = -\frac{e^2}{8\pi\varepsilon_0 u_B}\frac{1}{n^2}. \tag{3.107}$$

When electron energy changcs, the light with the frequency of

$$\nu_{n,n'} = \frac{E_{en} - E_{en'}}{h} \tag{3.108}$$

is absorbed or emitted so that the sum of the electron energy and the photon energy is conserved. This estimation is consistent with the experimental results shown in (3.104), taking the Rydberg constant,

$$Ry = \frac{\mu_e e^4}{8\varepsilon_0^2 h^3 c}. \tag{3.109}$$

Because of the discrete energy level, electrons can stay in the orbits.

3.8.2 The wave nature of particles

The idea of matter waves was given by de Broglie [19], indicating that all particles have a wave nature. The energy of each particle is $E = h\nu$, and the momentum is

$\vec{p} = h\vec{k}$ ($|\vec{k}| = 1/\lambda$), where ν is the frequency, $\vec{k}$ is the wavenumber vector, and λ is the wavelength (this relation was confirmed with light). From the relation $|\vec{p}| = h/\lambda$, (3.105) is interpreted as the description of the resonant wavelength regulated by the bounding state. The particles in a limited space area can take only discrete energy because the waves can take only resonant wavelengths (integer division of the size of the area), which determines the kinetic energy. Particles bound in a smaller area have a larger minimum energy because of a shorter wavelength (larger momentum), as shown in Figure 3.14.

The properties of all particles are given using a wave function $\mathbf{\Phi}$. The wave functions are often given using a complex function $\exp(ix) = e^{ix} = \cos x + i \sin x$, because the absolute value does not change with the oscillation ($|\exp(ix)| = 1$). The existence probability is given by $|\mathbf{\Phi}|^2$, in analogy with the fact that the photon density is proportional to the square of the oscillation amplitude of the electromagnetic field, as shown in (3.43). Physical value X is given by

$$\int \Phi^* \widetilde{X} \Phi d\vec{r} \tag{3.110}$$

as the average for the given state. Here, $\widetilde{X}$ is the operator for X. Wavefunctions having deterministic values of X (called eigenvalue X_e) are called eigenfunctions Φ_e, which satisfy

$$\check{X}\Phi_e = X_e \Phi_e. \tag{3.111}$$

All functions can be expressed as a linear combination of eigenfunctions as follows:

$$\begin{aligned}
&\Phi = \sum a_i \Phi_{ei} \\
&\int \Phi_{ei}^* \Phi_{ei} d\vec{r} = 1 \\
&\int \Phi_{ei}^* \Phi_{ej} d\vec{r} = 0 \, (i \neq j) \\
&\sum |a_i|^2 = 1.
\end{aligned} \tag{3.112}$$

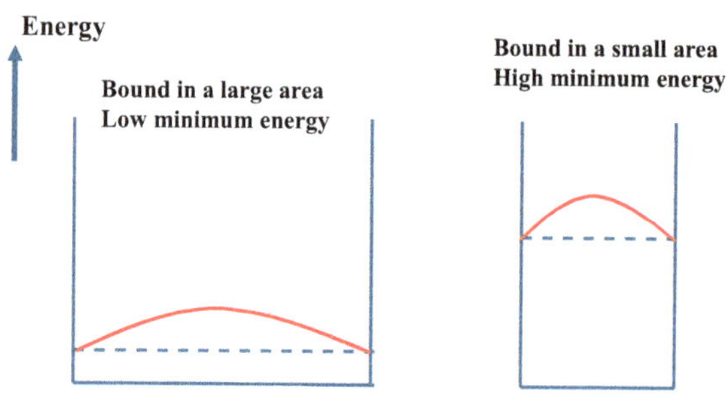

Figure 3.14 Particles bounded in a smaller area have a larger minimum energy because of a shorter wavelength

Before the measurement of X, this particle takes multiple physical values simultaneously. The measurement result of X is one of the eigenvalues X_{ei} with a probability of $|a_i|^2$. Owing to the measurement, the wavefunction changes to the eigenfunction corresponding to the measured eigenvalue. The change in the quantum state by the measurement is called "quantum destruction." Values of a_i are obtained by

$$a_i = \int \Phi_{ei}^* \Phi d\vec{r}. \quad (3.113)$$

3.8.3 The uncertainty principle

Here, we discuss the possibility of getting two deterministic physical values simultaneously. The measurement uncertainties of X and Y are given by $(X - X_e)^2$ and $(\check{Y} - Y_e)^2$. With the operators, we assume $\check{X}\check{Y} - \check{Y}\check{X} = \delta$. Using the real arbitral coefficients (dimensions of $C_{X,Y}$ (dimensions of $C_X X$ and $C_Y Y$ are unified)

$$\left[C_X(\check{X} - X_e) - iC_Y(\check{Y} - Y_e)\right]\left[C_X(\check{X} - X_e) + iC_Y(\check{Y} - Y_e)\right] > 0$$

$$C_X^2(\check{X} - X_e)^2 + C_Y^2(\check{Y} - Y_e)^2 + iC_X C_Y(\check{X}\check{Y} - \check{Y}\check{X}) > 0$$

$$\min\left[C_X^2(\Delta X)^2 + C_Y^2(\Delta Y)^2\right] = 2C_X C_Y \Delta X \Delta Y > C_X C_Y |\delta| \qquad \Delta X \Delta Y \geq |\delta|/2$$

$$\Delta X = |X - X_e|, \quad \Delta Y = |Y - Y_e| \quad (3.114)$$

Then X and Y cannot be deterministic simultaneously when $\delta \neq 0$ and the relation of uncertainties of both values is given by $\Delta X \Delta Y > |\delta|/2$.

With deterministic values of energy E and momentum $p_{x,y,z}$, the wavefunction is given by

$$\tilde{\Phi} = \exp\left[\frac{2\pi}{h} i(Et + p_x x + p_y y + p_z z)\right] \quad (3.115)$$

and

$$\widetilde{H}\Phi = \frac{h}{2\pi i}\frac{\partial \Phi}{\partial t} = E\Phi,$$

$$\widetilde{p_q} = \frac{h}{2\pi i}\frac{\partial \Phi}{\partial q} = p_q \Phi \quad q = x, y, z, \quad (3.116)$$

is derived (h: Planck constant). Here, $\widetilde{H}$ is the operator called Hamiltonian. Note

$$\widetilde{H}t - t\widetilde{H} = \widetilde{p_q}q - q\widetilde{p_q} = \frac{h}{2\pi i}. \quad (3.117)$$

and

$$\Delta E \Delta t > \frac{h}{4\pi}, \quad \Delta p_q \Delta q > \frac{h}{4\pi}. \quad (3.118)$$

is derived. The phase of the wavefunction given by Et/h and $p_q q/h$ has the uncertainty of the order of $\pm 1/2$ radian. Figure 3.14 investigates the result of the uncertainty of the position and momentum: with smaller position uncertainty by binding $0 < x < L$, the uncertainty of momentum between $p = \pm h/2L$ is increased, and the kinetic energy is increased. Here we consider a uniform traveling wave in the z-direction, $\Phi = \exp\left[\frac{2\pi i p_z z}{h}\right]$. With this wave, the momentum in the x- and y-directions is deterministically zero while the positions in both directions are totally random. When this wave passes through a slit with $0 < x < L$, the position in the x-direction is limited in this area, but the momentum uncertainty in the x-direction within $-\frac{h}{4\pi L} < p_x < \frac{h}{4\pi L}$ is induced by the diffraction, as shown in Figure 3.15.

The uncertainty principle was established without distinguishing between the limit of the measurement uncertainty and the quantum fluctuation. Ozawa insisted that these uncertainty components should be distinguished, and the uncertainty principle between measurement uncertainty between X and Y should be corrected to [20]

$$\Delta X \Delta Y + \Delta X \Delta Q_P + \Delta Y \Delta Q_X \geq \frac{h}{4\pi}$$
$$\Delta Q_{X,Y}: \text{ quantumfluctuationof } X \text{ and } Y. \tag{3.119}$$

Then the measurement uncertainty can be

$$\Delta X \Delta Y < \frac{h}{4\pi}, \tag{3.120}$$

which was experimentally confirmed by the measurement of the nuclear spin [21].

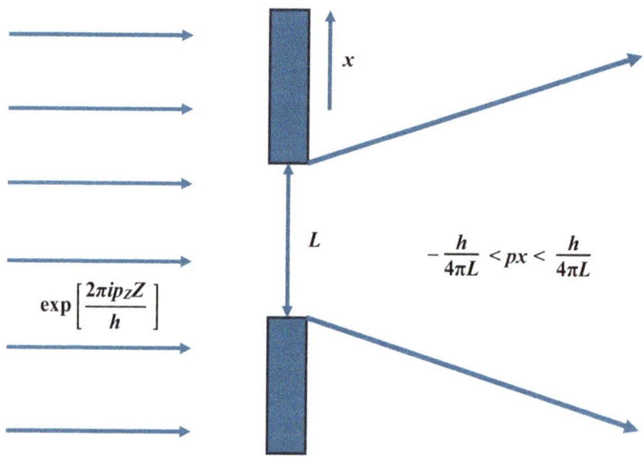

Figure 3.15 When a parallel wave passes through a slit, there is a diffraction. Reducing the position uncertainty in the x-direction by the slit, the uncertainty of the momentum is increased.

3.8.4 The Schrödinger equation

The Schroedinger equation for particles with mass m is derived from the relation between energy E and momentum $p_{x,y,z}$ given by Newtonian mechanics

$$E = \frac{p_x^2 + p_y^2 + p_z^2}{2m} + V(x,y,z)$$

$$V(x,y,z): \text{ potentialenergy}. \tag{3.121}$$

Using the operators of energy and momentum shown in (3.116), (3.121) is rewritten as

$$\widetilde{H}\Phi = \frac{h}{2\pi i}\frac{\partial\Phi}{\partial t} = \left[-\frac{h^2}{8\pi^2 m}\left(\frac{\partial^2}{\partial x^2} + \frac{\partial^2}{\partial y^2} + \frac{\partial^2}{\partial z^2}\right) + V(x,y,z)\right]\Phi. \tag{3.122}$$

When Φ is an energy eigenfunction, (3.122) is rewritten as

$$\Phi = \exp\left[\frac{2\pi i}{h}Et\right]\Psi(x,y,z)$$

$$E\Psi = \left[-\frac{h^2}{8\pi^2 m}\left(\frac{\partial^2}{\partial x^2} + \frac{\partial^2}{\partial y^2} + \frac{\partial^2}{\partial z^2}\right) + V(x,y,z)\right]\Psi. \tag{3.123}$$

When the potential is given by a spherically symmetric force (depending only $r = \sqrt{x^2 + y^2 + z^2}$), it is convenient to use the polar coordinate (r,θ,φ), $x = r\sin\theta\cos\varphi$, $y = r\sin\theta\sin\varphi, z = r\cos\theta$. Then (3.122) is rewritten as

$$\widetilde{H} = \left[-\frac{h^2}{8\pi^2 m}\left(\frac{\partial^2}{\partial r^2} + \frac{2}{r}\frac{\partial}{\partial r}\right) + \frac{1}{2mr^2}\widetilde{L}^2 + V(r)\right],$$

$$\breve{L}_x = \frac{h}{2\pi i}\left[-\sin\varphi\frac{\partial}{\partial\theta} - \frac{\cos\theta}{\sin\theta}\cos\varphi\frac{\partial}{\partial\varphi}\right],$$

$$\breve{L}_y = \frac{h}{2\pi i}\left[\cos\varphi\frac{\partial}{\partial\theta} - \frac{\cos\theta}{\sin\theta}\sin\varphi\frac{\partial}{\partial\varphi}\right],$$

$$\breve{L}_z = \frac{h}{2\pi i}\frac{\partial}{\partial\varphi},$$

$$\breve{L}^2 = \breve{L}_x^{\ 2} + \breve{L}_y^{\ 2} + \breve{L}_z^{\ 2} = -\left(\frac{h}{2\pi}\right)^2\left[\frac{\partial^2}{\partial\theta^2} + \frac{\cos\theta}{\sin\theta}\frac{\partial}{\partial\theta} + \frac{1}{(\sin\theta)^2}\frac{\partial^2}{\partial\varphi^2}\right], \tag{3.124}$$

where $\breve{L}_q$ is the operator of the angular momentum by the rotation around the q-axis ($q = x, y, z$). Here, we assume that the wave function is given by the formula of

$\Psi(r, \theta, \phi) = R(r)Y(\theta)\Theta(\varphi)$. The eigenfunction and eigenvalue of L_z are given by

$$\Theta(\varphi) = \frac{1}{\sqrt{2\pi}} \exp(iM_L\varphi)$$

$$L_z = \frac{h}{2\pi} M_L, \tag{3.125}$$

where M_L is an integer so that $\Theta(\varphi + 2\pi) = \Theta(\varphi)$ is satisfied. M_L is called the "magnetic quantum number." The absolute value of the angular momentum is given by a rotational quantum number L, and M_L is an integer between $\pm L$. We cannot determine the direction of the xy-axis because $|\Theta(\varphi)|^2$ is uniform against φ (uncertainty between φ and L_z), and $L_{x,y}$ cannot be determined except for when $L_x = L_y = L_z = 0$. This uncertainty relation between $L_{x,y,z}$ is also derived from

$$\check{L}_x\check{L}_y - \check{L}_y\check{L}_x = i\frac{h}{2\pi}\check{L}_z, \quad \check{L}_y\check{L}_z - \check{L}_z\check{L}_y = i\frac{h}{2\pi}\check{L}_x, \quad \check{L}_z\check{L}_x - \check{L}_x\check{L}_z = i\frac{h}{2\pi}\check{L}_y \tag{3.126}$$

Here, we consider operators

$$\check{L}_\pm = \check{L}_x \pm i\check{L}_y = \frac{h}{2\pi i}\exp(\pm i\varphi)\left[\pm i\frac{\partial}{\partial\theta} - \frac{\cos\theta}{\sin\theta}\frac{\partial}{\partial\varphi}\right] \tag{3.127}$$

Then

$$\check{L}_\pm \Theta_{M_L}(\varphi) \propto \exp[i(M_L \pm 1)\varphi] \propto \Theta_{M_L \pm 1}(\varphi) \tag{3.128}$$

and

$$\check{L}^2 = \check{L}_x^{\,2} + \check{L}_y^{\,2} + \check{L}_z^{\,2} = \check{L}_\mp\check{L}_\pm + \check{L}_z^{\,2} \pm \frac{h}{2\pi}\check{L}_z \tag{3.129}$$

are obtained. To prohibit the wave function of $M_L = \pm(L+1)$, there is a requirement

$$\check{L}_\pm Y_L^{\pm L}(\theta)\Theta_{\pm L}(\varphi) = 0$$

$$\check{L}^2 Y_L^{\pm L}(\theta)\Theta_{\pm L}(\varphi) = \left(\frac{h}{2\pi}\right)^2 L(L+1) Y_L^{\pm L}(\theta)\Theta_{\pm L}(\varphi) \tag{3.130}$$

The eigenvalue of the absolute value of the angular momentum is independent of M_L. The square of the absolute value of the angular momentum is $(h/2\pi)^2 L(L+1)$. The following solutions are obtained from (3.127) and (3.130)

$$\check{L}_\pm Y_L^{\pm L}(\theta)\Theta_{\pm L}(\varphi) = \frac{h}{2\pi i}\exp(\pm i\varphi)\left[\pm i\frac{\partial}{\partial\theta} - \frac{\cos\theta}{\sin\theta}\frac{\partial}{\partial\varphi}\right] Y_L^{\pm L}(\theta)\Theta_{\pm L}(\varphi) = 0$$

$$\pm i\frac{\partial}{\partial\theta} Y_L^{\pm L}(\theta) = \pm L\frac{\cos\theta}{\sin\theta} Y_L^{\pm L}(\theta)$$

$$\int \frac{dY_L^{\pm L}(\theta)}{Y_L^{\pm L}(\theta)} = L\int \frac{\cos\theta}{\sin\theta}d\theta = L\int \frac{d(\sin\theta)}{\sin\theta}$$

$$Y_L^{\pm L}\quad(\theta) \propto [\sin\theta]^L$$

$$Y_L^{\pm(L-i)}\quad(\theta) \propto \left(\check{L_\mp}\right)^i Y_L^{\pm L}\quad(\theta)$$

$$Y_L^{\pm(L-1)}\quad(\theta) \propto \check{L_\mp}\ Y_L^{\pm L}\quad(\theta) \propto [\sin\theta]^{L-1}\cos\theta$$

$$Y_0^o(\theta) = \sqrt{\frac{1}{2}},\quad Y_1^0(\theta) = \sqrt{\frac{3}{2}}\cos\theta,\quad Y_1^{\pm 1}(\theta) = \frac{\sqrt{3}}{2}\sin\theta,$$

$$Y_2^0(\theta) = \frac{1}{2}\sqrt{\frac{5}{2}}\left[3(\cos\theta)^2 - 1\right],\quad Y_2^{\pm 1}(\theta) = \frac{\sqrt{15}}{2}\sin\theta\cos\theta,\quad Y_2^{\pm 2}(\theta)$$
$$= \frac{\sqrt{15}}{4}(\sin\theta)^2. \tag{3.131}$$

$R(r)$ is obtained using

$$ER(r) = \left[-\frac{h^2}{8\pi^2 m}\left(\frac{\partial^2}{\partial r^2} + \frac{2}{r}\frac{\partial}{\partial r}\right) + \frac{h^2 L(L+1)}{8\pi^2\mu_e r^2} - V(r)\right]R(r). \tag{3.132}$$

Here, we consider the energy of an electron in a hydrogen atom, taking $m \to \mu_e = \frac{m_e m_p}{m_e + m_p}$ $m_{e,p}$: mass of electron and proton

$$V(r) \to -\frac{e^2}{4\pi\varepsilon_0 r}. \tag{3.133}$$

Equation (3.132) is rewritten as

$$ER(r) = \left[-\frac{h^2}{8\pi^2\mu_e}\left(\frac{\partial^2}{\partial r^2} + \frac{2}{r}\frac{\partial}{\partial r}\right) + \frac{h^2 L(L+1)}{8\pi^2\mu_e r^2} - \frac{e^2}{4\pi\varepsilon_0 r}\right]R(r)$$
$$\left[-\frac{\partial^2}{\partial r^2} - \frac{2}{r}\frac{\partial}{\partial r} + \frac{L(L+1)}{r^2} - \frac{2}{a_B r}\right]R(r) = \zeta R(r)$$

$a_B = \frac{\varepsilon_0 h^2}{\pi\mu_e e^2}$ (Bohr radius)

$$\zeta = \frac{8\pi^2\mu_e}{h^2}E_e. \tag{3.134}$$

We assume the formula

$$R(r) = G(r)\exp(-\alpha r). \tag{3.135}$$

Then (3.134) is rewritten as

$$\left[-\frac{\partial^2}{\partial r^2}G(r)+2\alpha\frac{\partial}{\partial r}G(r)-\frac{2}{r}\left(\frac{\partial}{\partial r}G(r)-\alpha G(r)\right)+\frac{L(L+1)}{r^2}G(r)-\frac{2}{a_B r}G(r)-\alpha^2 G(r)\right]=\zeta G(r). \tag{3.136}$$

Requiring (3.136) as an identity, the following relation is given.

$$\begin{aligned}
&-\alpha^2=\zeta\,E_e=-\frac{h^2}{8\pi^2\mu_e}\alpha^2\\
&\widetilde{H_r}G(r)=\widetilde{H_{r1}}\,G(r)+\widetilde{H_{r2}}\,G(r)=0\\
&\widetilde{H_{r1}}\,G(r)=2\left[\alpha\frac{\partial}{\partial r}G(r)+\frac{1}{r}\left(\alpha-\frac{1}{a_B}\right)G(r)\right]\\
&\widetilde{H_{r2}}\,G(r)=-\frac{\partial^2}{\partial r^2}G(r)-\frac{2}{r}\frac{\partial}{\partial r}G(r)+\frac{L(L+1)}{r^2}G(r).
\end{aligned} \tag{3.137}$$

We consider the limit of $r\to\infty$ and $r\to 0$. Note that $H_{r1}G(r)$ and $H_{r2}G(r)$ are functions of r with dimensions one and two orders lower than $G(r)$, respectively. With $r\to\infty$, $H_{r2}G(r)$ is negligibly small in comparison with $H_{r1}G(r)$. Therefore, $H_rG(r)=0$ is obtained by

$$\begin{aligned}
&\widetilde{H_r}G(r)\approx\widetilde{H_r}G(r)=0\\
&\frac{\partial}{\partial r}G(r)=\frac{1}{r}\left(\frac{1}{a_B\alpha}-1\right)G(r),\\
&\int\frac{dG(r)}{G(r)}=\left(\frac{1}{a_B\alpha}-1\right)\int\frac{dr}{r},\\
&\ln G(r)=\left(\frac{1}{a_B\alpha}-1\right)\ln r+const,\\
&G(r)\propto r^{n-1},\quad n=\frac{1}{a_B\alpha},\\
&E_e=-\frac{h^2}{8\pi^2\mu_e}\alpha^2=-\frac{h^2}{8\pi^2\mu_e}\left(\frac{1}{a_Bn}\right)^2=-\frac{e^2}{8\pi\varepsilon_0 a_B}\frac{1}{n^2}.
\end{aligned} \tag{3.138}$$

The energy eigenvalues shown in (3.138) agree with (3.107). From the following procedure, n (principal quantum number) is required to be an integer. With $r\to 0$,

$$\begin{aligned}
&\widetilde{H_r}G(r)\approx\widetilde{H_{r2}}\,G(r)=0\\
&-\frac{\partial^2}{\partial r^2}G(r)-\frac{2}{r}\frac{\partial}{\partial r}G(r)+\frac{L(L+1)}{r^2}G(r)=0.\\
&G(r)\propto r^L,\quad r^{-(L+1)}.
\end{aligned} \tag{3.139}$$

Requiring that $G(r)$ does not diverge at $r \to 0$, $G(r) \propto r^L$. L is required to be an integer as already shown in this subchapter. Here we consider the formula of $G_{n,L}(r) = \sum_{q=L}^{n-1} c_q r^q$ $(L \leq q < n)$ with the relation

$$c_q \widetilde{H_{r2}} r^q = c_q[-q(q+1) + L(L+1)]r^{q-2},$$
$$c_q \widetilde{H_{r1}} r^q = \frac{2}{a_B n} c_q (q - n + 1) r^{q-1}. \tag{3.140}$$

To satisfy $H_r G_{n,L}(r) = 0$, the following relation is required.

$$c_q \widetilde{H_{r2}} r^q + c_{q-1} \widetilde{H_{r1}} r^{q-1} = \left\{ c_q[-q(q+1) + L(L+1)] - \frac{2}{a_B n} c_{q-1}(n-q) \right\} r^{q-2} = 0$$
$$c_q[-q(q+1) + L(L+1)] = \frac{2}{a_B n} c_{q-1}(n-q). \tag{3.141}$$

Equation (3.141) requires $c_q = 0$ with $q < L$ and $q \geq n$; therefore, n must be an integer. The solutions of $R_{n,L}(r)$ with (n, L) states are given by

$$R_{1,0}(r) = 2\left(\frac{1}{a_B}\right)^{\frac{3}{2}} \exp\left(-\frac{r}{a_B}\right)$$
$$R_{2,1}(r) = \frac{1}{2\sqrt{6}}\left(\frac{1}{a_B}\right)^{\frac{3}{2}} \left(\frac{r}{a_B}\right) \exp\left(-\frac{r}{2a_B}\right)$$
$$R_{2,0}(r) \propto \frac{1}{\sqrt{2}}\left(\frac{1}{a_B}\right)^{\frac{3}{2}} \left(1 - \frac{r}{2a_B}\right) \exp\left(-\frac{r}{2a_B}\right). \tag{3.142}$$

The motion of the electron in the $L = 0$ state is interpreted as the vibrational motion in the radial direction because the distribution of the electron wavefunction is not zero at $r = 0$. The electron distribution in the $(n = 1, L = 0)$ state is derived from the uncertainty between the position and momentum in the radial direction. Taking $p_r = h/4\pi r$, the energy $(p_r^2/2m) - e^2/4\pi\varepsilon_0 r$ is minimum with $a_B/4$. The distribution at small r becomes smaller at higher L states, because of the centrifugal force. Energy eigenvalues derived from the Schrödinger equation are determined only by n. However, the actual energy structure is more complicated because of electron spin and relativistic effects, as shown in Chapter 3.9.2.

Next, we consider the energy eigenvalues of particles trapped in a harmonic potential, with which the Hamiltonian with the vibrational frequency of ν_{vib} is given by

$$\widetilde{H}\Phi = \left[\frac{\tilde{p}^2}{2m} + \frac{m}{2}(2\pi\nu_{vib})^2 x^2\right]\Phi \tag{3.143}$$

where m is the mass of the particle. For the eigenfunction corresponding to the energy eigenvalue of E_V, (3.143) is rewritten as

$$\Phi = \exp\left(\frac{2\pi i E_V}{h}t\right)\Psi$$

$$\left[\frac{\tilde{p}^2}{2m} + \frac{m}{2}(2\pi\nu_{vib})^2 x^2\right]\Psi = E_V\Psi \tag{3.144}$$

Here, we consider operators

$$\widetilde{a^{\pm}} = \sqrt{\frac{1}{2m}}\tilde{p} \pm i\sqrt{\frac{m}{2}}(2\pi\nu_{vib})x. \tag{3.145}$$

Then

$$\widetilde{H} = \widetilde{a^-}\widetilde{a^+} - \frac{h\nu_{vib}}{2} = \widetilde{a^+}\widetilde{a^-} + \frac{h\nu_{vib}}{2}$$

$$\widetilde{a^-}\widetilde{a^+} - \widetilde{a^+}\widetilde{a^-} = h\nu_{vib}, \tag{3.146}$$

is obtained using (3.117). When $\Psi(E_V)$ is the eigenfunction with the eigenenergy of E_V,

$$\widetilde{H}\left[\widetilde{a^{\pm}}\Psi(E_V)\right] = \left[\widetilde{a^{\mp}}\widetilde{a^{\pm}}\widetilde{a^{\mp}} \mp \frac{h\nu_{vib}}{2}\widetilde{a^{\pm}}\right]\Psi(E_V)$$

$$= \left[\left(\widetilde{a^{\pm}}\widetilde{a^{\mp}} \pm h\nu_{vib}\right)\widetilde{a^{\pm}} \mp \frac{h\nu_{vib}}{2}\widetilde{a^{\pm}}\right]\Psi(E_V)$$

$$= \widetilde{a^{\pm}}\left[\left(\widetilde{a^{\mp}}\widetilde{a^{\pm}} \mp \frac{h\nu_{vib}}{2}\right) \pm h\nu_{vib}\right]\Psi(E_V)$$

$$= \widetilde{a^{\pm}}(H \pm h\nu_{vib})\Psi(E_V)$$

$$= (E_V \pm h\nu_{vib})\left(\widetilde{a^{\pm}}\Psi(E_V)\right), \tag{3.147}$$

and $\left(\widetilde{a^{\pm}}\Psi(E_V)\right) \propto \Psi(E_V \pm h\nu_{vib})$ is derived.

We can also give another interpretation of (3.147). From the constancy of the total energy E_V, the following expression is given

$$\sqrt{\frac{1}{2m}}\tilde{p} = \sqrt{E_V}\cos(2\pi\nu_{vib}t), \quad \sqrt{\frac{m}{2}}(2\pi\nu_{vib})x = \sqrt{E_V}\sin(2\pi\nu_{vib}t)$$

$$\widetilde{a^{\pm}} = \sqrt{E_V}[\cos(2\pi\nu_{vib}t) \pm i\sin(2\pi\nu_{vib}t)] = \sqrt{E_V}\exp(\pm 2\pi i\nu_{vib}t)$$

$$\left(\widetilde{a^{\pm}}\Phi(E_V)\right) \propto \sqrt{E_V}\exp\left[\frac{2\pi i}{h}(E_V \pm h\nu_{vib})\right] \propto \Phi(E_V \pm h\nu_{vib}). \tag{3.148}$$

To prohibit the negative value of eigen energy, the eigenfunction with the minimum energy eigenvalue Ψ_0 must hold $a^-\Psi_0 = 0$, and the minimum energyeigen value is

$$E_V(0) = \frac{h\nu_{vib}}{2}. \tag{3.149}$$

$\Psi(0)$ is obtained by

$$a^{-}\Psi(0) = 0$$

$$\frac{\partial}{\partial x}\Psi(0) = -\frac{4\pi^2 m\nu_{vib}}{h}x\Psi(0)$$

$$\int \frac{1}{\Psi(0)}d\Psi(0) = -\int \frac{4\pi^2 m\nu_{vib}}{h}xdx$$

$$\Psi(0) \propto \exp\left(-\frac{2\pi^2 m\nu_{vib}}{h}x^2\right). \tag{3.150}$$

The eigenfunctions with higher eigenenergy are given by

$$\widetilde{a^{\pm}}\Psi(n_{vib}) \propto \Psi(n_{vib} \pm 1)$$

$$\Psi(n_{vib}) \propto \left(\widetilde{a^{+}}\right)^{n_{vib}}\Psi(0), \tag{3.151}$$

where n_{vib} is an integer. For example,

$$\Psi(1) \propto \text{xexp}\left(-\frac{2\pi^2 m\nu_{vib}}{h}x^2\right). \tag{3.152}$$

The energy eigenvalues are given by

$$E_V(n_{vib}) = \left(n_{vib} + \frac{1}{2}\right)h\nu_{vib}. \tag{3.153}$$

The minimum value of the vibrational energy is also derived from the uncertainty principle between p and x.

$$\frac{p^2}{2m} + \frac{m(2\pi\nu_{vib})^2 x^2}{2} \geq (2\pi\nu_{vib})(px) \geq \frac{h\nu_{vib}}{2}. \tag{3.154}$$

3.8.5 Spin

The quantum energy state of an electron in an atom is given by the principal quantum number n, rotational quantum number L, and magnetic quantum number M_L, as shown in Chapter 3.8.4. An electron is a fermion (Chapter 7.2), and only one electron can be in a quantum state. However, two electrons can be in an (n, L, M_L) state, which means there are two states with electrons. The deflection in two opposite directions with the same angle was observed with an Ag atomic beam passing through an area with an inhomogeneous magnetic field [22]. The electron orbital angular momentum of the Ag atom was zero. This result shows that the electron has two states, like a permanent magnet: the S-pole or N-pole in the direction of the magnetic field. These two states of electrons were described as the virtual angular momentum states ($S = 1/2, M_S = \pm 1/2$) called spin. The spin state is a property of the electron itself, and the eigenfunction is not derived from the density distribution in a potential field. Pauli proposed describing two spin states using a two-dimensional vector [23]. The eigenfunction of each spin state is as follows

$$M_S = \frac{1}{2} \rightarrow \xi_+ = \begin{pmatrix} 1 \\ 0 \end{pmatrix} M_S = -\frac{1}{2} \rightarrow \xi_- = \begin{pmatrix} 0 \\ 1 \end{pmatrix}. \tag{3.155}$$

The operators of the spin components in the x, y, and z directions are

$$\widetilde{S_{1/2q}} = \frac{h}{4\pi}\sigma_q \; q = x, y, z$$

$$\sigma_{1/2x} = \begin{pmatrix} 0 & 1 \\ 1 & 0 \end{pmatrix}, \quad \sigma_{1/2y} = \begin{pmatrix} 0 & -i \\ i & 0 \end{pmatrix}, \quad \sigma_{1/2z} = \begin{pmatrix} 1 & 0 \\ 0 & -1 \end{pmatrix}, \tag{3.156}$$

where $\sigma_{1/2q}$ are the Pauli matrices. The following relations are satisfied with Pauli matrices and spin operators.

$$\sigma_{1/2x}^2 = \sigma_{1/2y}^2 = \sigma_{1/2z}^2 = I \; I = \begin{pmatrix} 1 & 0 \\ 0 & 1 \end{pmatrix} \rightarrow S_{1/2x}^2 + S_{1/2y}^2 + S_{1/2z}^2$$
$$= \frac{3}{4}\left(\frac{h}{2\pi}\right)^2 I \tag{3.157}$$

$$\sigma_{1/2x}\sigma_{1/2y} + \sigma_{1/2y}\sigma_{x1/2} = \sigma_{1/2x}\sigma_{1/2z} + \sigma_{1/2z}\sigma_{1/2x} = \sigma_{1/2y}\sigma_{1/2z} + \sigma_{1/2z}\sigma_{1/2y} = 0,$$
$$\sigma_{1/2x}\sigma_{1/2y} - \sigma_{1/2y}\sigma_{1/2x} = 2i\sigma_{1/2z} \rightarrow S_{1/2x}S_{1/2y} - S_{1/2y}S_{1/2x} = i\frac{h}{2\pi}S_{1/2z},$$
$$\sigma_{1/2y}\sigma_{1/2z} - \sigma_{1/2z}\sigma_{1/2y} = 2i\sigma_{1/2x} \rightarrow S_{1/2y}S_{1/2z} - S_{1/2z}S_{1/2y} = i\frac{h}{2\pi}S_{1/2x},$$
$$\sigma_{1/2z}\sigma_{1/2x} - \sigma_{1/2x}\sigma_{1/2z} = 2i\sigma_{1/2y} \rightarrow S_{1/2z}S_{1/2x} - S_{1/2x}S_{1/2z} = i\frac{h}{2\pi}S_{1/2y} \tag{3.158}$$

Equation (3.157) corresponds to (3.130) substituting $L_{x,y,z} \rightarrow S_{x,y,z}$. Equation (3.158) corresponds to (3.126). These relations indicate that spin has the characteristic of angular momentum. The wavefunction, including the electron spin, should be given by vectors, which cannot be derived by the Schrödinger equation, which is scalar.

The existence of electron spin $\vec{S}$ gives the energy splitting (called fine structure) by coupling with the orbital angular momentum $\vec{L}$, which is described by $\vec{J} = \vec{L} + \vec{S}$. For an electron ($S = 1/2$), the $J = L + 1/2$ and $J = L - 1/2$ states denote the states in which the orbital angular momentum and spin are parallel and antiparallel, respectively. We can imagine the interaction between the magnetic field induced by the revolution motion of the electron and the electron spin as a permanent magnet.

The quantum energy state is given by (n, L, S, J) with the description of $n^{2S+1}X_J$, where X = S $(L=0)$, P $(L=1)$, and D $(L=2)$. For example, $1^2S_{1/2}$ denotes $\left(n=1, L=0, S=\frac{1}{2}, J=1/2\right)$, and $2^2P_{3/2}$ denotes $\left(n=2, L=1, S=\frac{1}{2}, J=3/2\right)$.

There are also particles with $S=1$. With this case, the wavefunction is given by

$$M_S = 1 \rightarrow \xi_1 = \begin{pmatrix} 1 \\ 0 \\ 0 \end{pmatrix} \quad M_S = 0 \rightarrow \xi_0 = \begin{pmatrix} 0 \\ 1 \\ 0 \end{pmatrix}$$

$$M_S = -1 \rightarrow \xi_{-1} = \begin{pmatrix} 0 \\ 0 \\ 1 \end{pmatrix}. \tag{3.159}$$

The operator is given by

$$\widetilde{S_{1q}} = \frac{h}{2\pi}\sigma_{1q} \; q = x, y, z$$

$$\sigma_{1x} = \frac{1}{\sqrt{2}}\begin{bmatrix} 0 & 1 & 0 \\ 1 & 0 & 1 \\ 0 & 1 & 0 \end{bmatrix} \quad \sigma_{1y} = \frac{1}{\sqrt{2}}\begin{bmatrix} 0 & -i & 0 \\ i & 0 & -i \\ 0 & i & 0 \end{bmatrix} \quad \sigma_{1z} = \begin{bmatrix} 1 & 0 & 0 \\ 0 & 0 & 0 \\ 0 & 0 & -1 \end{bmatrix}. \tag{3.160}$$

With this operator, the following relation holds, and the characteristic of the angular momentum also holds with $S=1$.

$$S_{1x}^2 + S_{1y}^2 + S_{1z}^2 = 2\left(\frac{h}{2\pi}\right)^2 I_1, \quad I_1 = \begin{pmatrix} 1 & 0 & 0 \\ 0 & 1 & 0 \\ 0 & 0 & 1 \end{pmatrix}$$

$$\begin{aligned} S_{1x}S_{1y} - S_{1y}S_{1x} &= i\frac{h}{2\pi}S_{1z}, \\ S_{1y}S_{1z} - S_{1z}S_{1y} &= i\frac{h}{2\pi}S_{1x}, \\ S_{1z}S_{1x} - S_{1x}S_{1z} &= i\frac{h}{2\pi}S_{1y} \end{aligned} \tag{3.161}$$

However, note that $\sigma_{1x}^2 = \sigma_{1y}^2 = \sigma_{1z}^2 = I_1$ is not satisfied.

3.9 Relativistic Quantum Mechanics

3.9.1 The Klein–Gordon equation

The Schrödinger equation is based on the relationship between the frequency (energy) and wavenumber (momentum) given in classical mechanics. Based on the theory of special relativity, the relationship between energy and momentum is given by (Equation (3.93)):

$$E^2 = \left(mc^2\right)^2 + c^2|\vec{p}|^2. \tag{3.162}$$

Using the energy and momentum operators shown in (3.116), the Klein–Gordon equation is given as [24]

$$-\left(\frac{h}{2\pi}\right)^2 \frac{\partial^2}{\partial t^2} = -\left(\frac{h}{2\pi}\right)^2 c^2 \left[\frac{\partial^2}{\partial x^2} + \frac{\partial^2}{\partial y^2} + \frac{\partial^2}{\partial z^2}\right] + \left(mc^2\right)^2 \tag{3.163}$$

is derived. However, the utility of the solutions is questionable because they derive the square of energy (not energy itself). The main problem is that negative probability and energy values are included in the solutions. The wave function is treated as a scalar; therefore, the effect of electron spin was not derived.

The Klein–Gordon equation was forgotten for a while. However, Pauli and Weisskopf determined that this equation is useful for treating boson particles with zero spin (called scalar particles), for which the wave function is treated as a scalar [25].

3.9.2 The Dirac equation

Dirac derived a Hamiltonian with a relativistic treatment [26]

$$\widetilde{H} = c\left[\widetilde{\alpha}_x \widetilde{p}_x + \widetilde{\alpha}_y \widetilde{p}_y + \widetilde{\alpha}_z \widetilde{p}_z\right] + \widetilde{\beta} mc^2 \tag{3.164}$$

with the requirement that $\widetilde{H}^2$ corresponds to the Klein–Gordon equation. For operators $\widetilde{\alpha_q}$ ($q = x, y, z$) and $\widetilde{\beta}$, the following relations must hold.

$$\begin{aligned} &\widetilde{\alpha_q}^2 = \widetilde{\beta}^2 = 1 \\ &\widetilde{\alpha}_x \widetilde{\alpha}_y + \widetilde{\alpha}_y \widetilde{\alpha}_x = \widetilde{\alpha}_y \widetilde{\alpha}_z + \widetilde{\alpha}_z \widetilde{\alpha}_y = \widetilde{\alpha}_z \widetilde{\alpha}_x + \widetilde{\alpha}_x \widetilde{\alpha}_z = 0 \\ &\widetilde{\alpha_q}\widetilde{\beta} + \widetilde{\beta}\widetilde{\alpha_q} = 0. \end{aligned} \tag{3.165}$$

Equation (3.165) cannot be satisfied with a scalar but can be satisfied using 4×4 matrices

$$\widetilde{\alpha_q} = \begin{pmatrix} 0 & \sigma_{1/2q} \\ \sigma_{1/2q} & 0 \end{pmatrix}, \quad \widetilde{\beta} = \begin{pmatrix} I & 0 \\ 0 & -I \end{pmatrix}, \tag{3.166}$$

where $\sigma_{1/2q}$ are the Pauli matrices (see (3.156)). The eigenfunctions are given by four-dimensional vectors.

$$\vec{\Psi} = \begin{pmatrix} \vec{u} \\ \vec{w} \end{pmatrix} \vec{u} = \begin{pmatrix} u_1 \\ u_2 \end{pmatrix} \vec{w} = \begin{pmatrix} w_1 \\ w_2 \end{pmatrix}. \tag{3.167}$$

First, we consider $\vec{p} = 0$ for simplicity. Then

$$\widetilde{H} = \begin{pmatrix} mc^2 & 0 & 0 & 0 \\ 0 & mc^2 & 0 & 0 \\ 0 & 0 & -mc^2 & 0 \\ 0 & 0 & 0 & -mc^2 \end{pmatrix} \tag{3.168}$$

and four eigenfunctions

$$\vec{\Psi} = \begin{pmatrix} 1 \\ 0 \\ 0 \\ 0 \end{pmatrix}, \begin{pmatrix} 0 \\ 1 \\ 0 \\ 0 \end{pmatrix}, \begin{pmatrix} 0 \\ 0 \\ 1 \\ 0 \end{pmatrix}, \begin{pmatrix} 0 \\ 0 \\ 0 \\ 1 \end{pmatrix} \tag{3.169}$$

are obtained. The energy eigenvalues are mc^2 for two solutions and $-mc^2$ for the other two solutions. Two solutions for each energy eigenvalue are interpreted as two spin states $\left(S = 1/2,\ M_S = \pm\frac{1}{2}\right)$. Applying a magnetic field, the energy shifts (called Zeeman energy shift) in opposite directions are induced for both spin states, as shown below. The meaning of negative rest energy is discussed in Chapter 8.1. The Dirac equation is valid for particles with $S = 1/2$. If we consider $S = 1$ using 6×6 matrices

$$\widetilde{\alpha_{q1}} = \begin{pmatrix} 0 & \sigma_{1q} \\ \sigma_{1q} & 0 \end{pmatrix}, \quad \widetilde{\beta_1} = \begin{pmatrix} I_1 & 0 \\ 0 & -I_1 \end{pmatrix}, \quad \widetilde{I_1} = \begin{pmatrix} 1 & 0 & 0 \\ 0 & 1 & 0 \\ 0 & 0 & 1 \end{pmatrix}, \tag{3.170}$$

Equation (3.165) is not satisfied. Therefore, the Dirac equation is not applicable for particles with $S = 1$.

Here, we consider the kinetic energy K with $E \approx mc^2$.

$$E\vec{u} = mc^2\vec{u} + c\left[\widetilde{\sigma_x}\widetilde{p_x} + \widetilde{\sigma_y}\widetilde{p_y} + \widetilde{\sigma_z}\widetilde{p_z}\right]\vec{w}$$

$$E\vec{w} = -mc^2\vec{w} + c\left[\widetilde{\sigma_x}\widetilde{p_x} + \widetilde{\sigma_y}\widetilde{p_y} + \widetilde{\sigma_z}\widetilde{p_z}\right]\vec{u}. \tag{3.171}$$

For the limit $K = E - mc^2 \ll mc^2$, (3.171) is rewritten as

$$K\vec{u} = c\left[\sigma_x\widetilde{p_x} + \sigma_y\widetilde{p_y} + \sigma_z\widetilde{p_z}\right]\vec{w}$$

using the approximation $(E + mc^2)\vec{w} = 2mc^2\vec{w}$

$$2mc^2\vec{w} = c\left[\sigma_x\widetilde{p_x} + \sigma_y\widetilde{p_y} + \sigma_z\widetilde{p_z}\right]\vec{u} \rightarrow \vec{w} = \frac{\left[\sigma_x\widetilde{p_x} + \sigma_y\widetilde{p_y} + \sigma_z\widetilde{p_z}\right]}{2mc}\vec{u}$$

based on (3.165)

$$K\vec{u} = \frac{\widetilde{p_x}^2 + \widetilde{p_y}^2 + \widetilde{p_z}^2}{2m}\vec{u} \tag{3.172}$$

Equation (3.172) corresponds to the Schrödinger equation.

Here we consider the energy shift induced by the magnetic field (called the Zeeman shift). For an electromagnetic field, we use the exchange $E \rightarrow E - q_e\Phi_e$

and $\vec{p} \rightarrow \vec{p} + q_e\vec{A}$ as shown in Chapter 3.2.2. We obtain $E' = E - mc^2$ using the classical approximation $(E + mc^2 \approx 2mc^2)$ for $\vec{A} = \frac{1}{2}(-yB_m, xB_m, 0)$ and $\overrightarrow{B_m} = (0, 0, B_m)$.

$$\left[\sigma_x\left(\frac{h}{2\pi i}\frac{\partial}{\partial x} - \frac{q_e y B_m}{2}\right)\right]^2 =$$

$$\begin{pmatrix} -\left(\frac{h}{2\pi}\right)^2\frac{\partial^2}{\partial x^2} - q_c B_m y\left(\frac{h}{2\pi i}\right)\frac{\partial}{\partial x} + \left(\frac{q_e y B_m}{2}\right)^2 & 0 \\ 0 & -\left(\frac{h}{2\pi}\right)^2\frac{\partial^2}{\partial x^2} - q_c B_m y\left(\frac{h}{2\pi i}\right)\frac{\partial}{\partial x} + \left(\frac{q_e y B_m}{2}\right)^2 \end{pmatrix}$$

$$\left[\sigma_y\left(\frac{h}{2\pi i}\frac{\partial}{\partial y} + \frac{q_e x B_m}{2}\right)\right]^2 =$$

$$\begin{pmatrix} -\left(\frac{h}{2\pi}\right)^2\frac{\partial^2}{\partial y^2} + q_c B_m x\left(\frac{h}{2\pi i}\right)\frac{\partial}{\partial y} + \left(\frac{q_e x B_m}{2}\right)^2 & 0 \\ 0 & -\left(\frac{h}{2\pi}\right)^2\frac{\partial^2}{\partial y^2} + q_c B_m x\left(\frac{h}{2\pi i}\right)\frac{\partial}{\partial y} + \left(\frac{q_e x B_m}{2}\right)^2 \end{pmatrix}$$

$$\sigma_x\left(\frac{h}{2\pi i}\frac{\partial}{\partial x} - \frac{q_e y B_m}{2}\right)\sigma_y\left(\frac{h}{2\pi i}\frac{\partial}{\partial y} + \frac{q_e x B_m}{2}\right) + \sigma_y\left(\frac{h}{2\pi i}\frac{\partial}{\partial y} + \frac{q_e x B_m}{2}\right)\sigma_x\left(\frac{h}{2\pi i}\frac{\partial}{\partial x} - \frac{q_e y B_m}{2}\right)$$

$$= \frac{q_e h}{2\pi}B_m\begin{pmatrix} 1 & 0 \\ 0 & -1 \end{pmatrix}. \qquad (3.173)$$

Taking

$$L_z = \frac{h}{2\pi}M_L = \left[x\left(\frac{h}{2\pi i}\frac{\partial}{\partial y}\right) - y\left(\frac{h}{2\pi i}\frac{\partial}{\partial x}\right)\right]$$

$$S_z = \frac{h}{2\pi}M_S == \frac{h}{4\pi}\sigma_z, \qquad (3.174)$$

$$E'\vec{u} = \left[\frac{\widetilde{p_x}^2 + \widetilde{p_y}^2 + \widetilde{p_z}^2}{2m} + \frac{q_e}{2m}L_z B_m + q_c\Phi\right]I\vec{u} + 2\frac{q_e}{2m}B_m S_z\vec{u}. \qquad (3.175)$$

is obtained. For an applied magnetic field, there are Zeeman energy shifts for both spin states in positive and negative directions with the same absolute value. For an electron, the Zeeman energy shift is described by

$$E_Z = \mu_B[M_L + g_S M_S]B$$

$$\mu_B = \frac{eh}{4\pi m_e}. \qquad (3.176)$$

The g-factor of the electron spin g_S is obtained as exactly 2 from (3.175), but it was experimentally determined to be 2.0023 [27]. This discrepancy is the "anomalous magnetic moment," which is derived based on quantum electrodynamics [28], taking the energy fluctuations in a vacuum into account. The uncertainties of experimental and calculative values of the anomalous magnetic moment have been reduced. Previously, no serious discrepancy had been discovered between the two results.

The existence of the Zeeman energy shift induced by the electron spin is not a relativistic effect, but it was obtained from the Dirac equation, which requires the Hamiltonian to be treated as matrices.

Next, m we consider the relativistic effect without an electromagnetic field. In (3.172), the approximation $(E + mc^2)\vec{w} = 2mc^2\vec{w}$ was used. For more detailed analysis, we consider with $(E + mc^2)\vec{w} = (2mc^2 + E')\vec{w}$ as follows.

$$\left(2mc^2 + E'\right)\vec{w} = c\left[\widetilde{\sigma_x p_x} + \widetilde{\sigma}_y\widetilde{p}_y + \widetilde{\sigma}_z\widetilde{p}_z\right]\vec{u}$$

$$\vec{w} = \frac{c\left[\widetilde{\sigma_x p_x} + \widetilde{\sigma}_y\widetilde{p}_y + \widetilde{\sigma}_z\widetilde{p}_z\right]\vec{u}}{2mc^2 + E'}$$

$$= \frac{c\left[\widetilde{\sigma_x p_x} + \widetilde{\sigma}_y\widetilde{p}_y + \widetilde{\sigma}_z\widetilde{p}_z\right]\vec{u}}{2mc^2} - \frac{E'c\left[\widetilde{\sigma_x p_x} + \widetilde{\sigma}_y\widetilde{p}_y + \widetilde{\sigma}_z\widetilde{p}_z\right]\vec{u}}{\left(2mc^2\right)^2}$$

$$E'\vec{u} = E'_{cl}\vec{u} + E'_{rel}\vec{u}$$

$$\frac{E'_{cl} = \widetilde{p_x}^2 + \widetilde{p_y}^2 + \widetilde{p_z}^2}{2m}$$

$$E'_{rel} = -c\left[\widetilde{\sigma_x p_x} + \widetilde{\sigma}_y\widetilde{p}_y + \widetilde{\sigma}_z\widetilde{p}_z\right]\frac{E'c\left[\widetilde{\sigma_x p_x} + \widetilde{\sigma}_y\widetilde{p}_y + \widetilde{\sigma}_z\widetilde{p}_z\right]}{\left(2mc^2\right)^2}. \tag{3.177}$$

To calculate the relativistic correction, E'_{rel} was calculated by taking $E' = E'_{cl}$ for simplicity and

$$E'_{rel} = -c\left[\widetilde{\sigma_x p_x} + \widetilde{\sigma}_y\widetilde{p}_y + \widetilde{\sigma}_z\widetilde{p}_z\right]\frac{\widetilde{p_x}^2 + \widetilde{p_y}^2 + \widetilde{p_z}^2}{2m}\frac{c\left[\widetilde{\sigma_x p_x} + \widetilde{\sigma}_y\widetilde{p}_y + \widetilde{\sigma}_z\widetilde{p}_z\right]\vec{u}}{\left(2mc^2\right)^2}$$

$$= -\frac{\left(\widetilde{p_x}^2 + \widetilde{p_y}^2 + \widetilde{p_z}^2\right)^2}{8m^3c^2}. \tag{3.178}$$

This term is derived as the third term of

$$E = mc^2\sqrt{1 + \left(\frac{|\vec{p}|}{mc}\right)^2} = mc^2 + \frac{|\vec{p}|^2}{2m} - \frac{|\vec{p}|^4}{8m^3c^2}. \tag{3.179}$$

The energy eigenvalue of the electron in a hydrogen atom or hydrogen-like ions (one electron and the nucleus with a charge of +Ze) is obtained strictly from the Dirac equation as

$$E_{n,J} = \frac{m_e c^2}{\sqrt{1 + \left(\frac{Z\alpha}{n - J - \frac{1}{2} + \sqrt{\left(J + \frac{1}{2}\right)^2 - (Z\alpha)^2}}\right)^2}}$$

$$\alpha = \frac{e^2}{2\varepsilon_0 hc} \text{ (fine structure constant)}. \tag{3.180}$$

For the value of $\alpha = 0.00116$, the following approximation is valid for a small Z.

$$\sqrt{\left(J+\frac{1}{2}\right)^2-(Z\alpha)^2}=\left(J+\frac{1}{2}\right)-\frac{(Z\alpha)^2}{2\left(J+\frac{1}{2}\right)}$$

$$E_{n,J}=\frac{m_ec^2}{\sqrt{1+\left(\frac{Z\alpha}{n-\frac{(Z\alpha)^2}{2\left(J+\frac{1}{2}\right)}}\right)^2}}=m_ec^2-\frac{m_ec^2Z^2\alpha^2}{2\left(n-\frac{(Z\alpha)^2}{2\left(J+\frac{1}{2}\right)}\right)^2}$$

$$\text{taking } \frac{1}{\left(n-\frac{(Z\alpha)^2}{2\left(J+\frac{1}{2}\right)}\right)^2}=\frac{1}{n^2}+\frac{(Z\alpha)^2}{n^3\left(J+\frac{1}{2}\right)}$$

$$=m_ec^2-\frac{m_eZ^2\alpha^2}{2n^2}--\frac{m_eZ^4}{2n^3}\frac{\alpha^4}{\left(J+\frac{1}{2}\right)}. \tag{3.181}$$

The first term denotes the rest energy of an electron. The second term represents the energy eigenvalue obtained using the Schrödinger equation, and the third term denotes the relativistic effect depending on J. Note that the relative motion between the electron and nucleus is not separated from the motion of the center of mass with the relativistic treatment (Chapter 3.6). Therefore, it is often treated by changing m_e to μ_e (reduced mass) in the second and third terms of (3.181), assuming that the relativistic effect due to the motion of the center of mass is negligible.

From the Schrödinger equation, the eigenenergy of an electron in a hydrogen atom is determined only by n. From the Dirac equation, an energy gap between the $2^2P_{3/2}$ and $2^2P_{1/2}$ states with a transition frequency of 10.9 GHz is derived. There should be no energy gap between $2^2S_{1/2}$ and $2^2P_{1/2}$ states from the Dirac equation, but there is a slight energy gap (transition frequency 1.05 GHz), called the Lamb shift [29]. The discovery of this slight energy gap played a significant role in developing the field of "quantum electrodynamics," considering vacuum energy fluctuations [28]. The energy in the $1^2S_{1/2}$ state is also shifted from the solution of the Dirac equation by the quantum electrodynamical effect. There is currently no significant discrepancy between measurement and calculation, taking the quantum electrodynamical effect into account.

The nucleus also has a spin, and the interaction between the electron spin and the electron orbital angular momentum induces another energy splitting, called hyperfine splitting. The transition frequency of the hyperfine splitting of hydrogen atoms in the $^1S_{1/2}$ state is 1.4 GHz.

The Dirac equation using a 4×4 matrix is applicable for the particles with $S = \frac{1}{2}$, and the Klein–Gordon equation is applicable for the particles with $S = 0$. For the particles with other spin values, the following equations are applicable.

The Rarita–Schwiner equation for massive particles with $S = \frac{3}{2}$ [30]

The Bargmann–Wigner equation for free particles with arbitral spin [31]

The Bargmann–Wigner equation is a set of linear partial differential equations with $2S$ components. The calculation of each component was based on the Dirac equation. However, it is difficult to use the Bargmann–Wigner equation for particles in an electromagnetic field.

References

[1] Hurai P. G. *Maxwell's Equations*. IEEE Press 2010; p 22.

[2] Fizeau H. 'Sur une expérience relative à la vitesse de propagation de la lumière'. *Comptes Rendus Hedomadaires des Seances de l'Academie des Science*; 1849; 29 90–92. https://www.academie-sciences.fr/pdf/dossiers/Fizeau/Fizeau_pdf/CR1849_p90.pdf.

[3] Foucault L. 'Determination experimentale de la vitesse de la lumière'. *Comptes Rendus Hedomadaires des Seances de l'Academie des Science*; 1852; 55 501–503.

[4] Foucault L. 'Sur les vitesses relatives de la lumier dans l'air et dans l'eau'. *Annales de Chemie et de Physique* [3], 1853; t.41 129–164.

[5] Michelson A. A. 'The relative motion of the earth and of the luminiferous ether'. *American Journal of Science*. 1881; 22 120–129. DOI:10.2475/ajs.s3-22.128.120.

[6] Brillet A. and Hall J. L. 'Improved laser test of the isotropy of space'. *Phys. Rev. Lett.* 1979; 42 549–552. DOI:10.1103/PhysRevLett.42.549.

[7] London R. *The Quantum Theory of Light* (third ed.), Cambridge Univ., 1973.

[8] Hertz H. R. 'Ueber einen Einfluss des ultravioletten Lichtes auf die electrische Entladung'. *Annalen der Physik*, 1887; 267 983–1000. DOI:10.1002/andp.18872670827.

[9] Einstein A. 'Über einen die Erzeugung und Verwandlung des Lichtes betreffenden heuristischen Gesichtspunkt'. *Annalen der Physik Ser 4*, 1905; 322 132–148. DOI: 10.1002/andp.19053220607.

[10] Compton A. H. 'A quantum theory of the scattering of X-rays by light elements'. *Phys. Rev.*, 1922; 21 483–502. DOI:10.1103/PhysRev.21.483.

[11] Einstein A. 'Relativity: The special and general theory', https://www.gutenberg.org/files/30155/30155-h/30155-h.htm; 1916 (translated 1920, New York: H. Holt and Compa).

[12] Dyson F. W., Eddington A. S. and Davidson C. 'A determination of the deflection of light by the sun's gravitational field, from observations made at the total eclipse of May 29, 1919'. *Philosophical Transactions of Royal Society*, 1920; 220A291–333. DOI:10.1098/rsta.1920.0009.

[13] The Event Horizon Telescope Collaboration. 'First M87 event horizon telescope results. I. The shadow of the supermassive black hole'. *Astrophys. J. Lett.*, 2019; 875 L1 1–28. DOI:10.3847/2041-8213/ab0ec7.

[14] Maccarone T. J., Kundu A., Zepf S. E. and Rhode K. L. 'A black hole in a globular cluster'. *Nature*, 2007; 445 183–185. DOI:10.1038/nature05434.

[15] Nobili A. and Will C. 'The real value of Mercury's perihelion advance'. *Nature*, 1986; 320 39–41. DOI:10.1038/320039a0.
[16] Balmer J. J. 'Notiz über die Spectrallinien des Wasserstoffs'. *Annahlen der Physik*, 1885; 261 80 DOI: 10.1002/andp.18852610506.
[17] Lyman T. 'The spectrum of hydrogen in the region of extremely short wave-length'. *Memoirs of the American Academy of Arts and Sciences, New Series*, 1906; 13 (3): 125–146.
[18] Bohr N. 'On the constitution of atoms and molecules'. *Phil. Mag.* Series 6, 1913; 26 1–24 DOI:10.1080/14786441308634955.
[19] de Broglie L. V. 'Recherches sur la théorie des quanta'. *Thesis, Paris,1924 Anne de Physique*, 1925; (10) 3 22.
[20] Ozawa M. 'Universally valid reformulation of the Heisenberg uncertainty principle on noise and disturbance in measurement'. *Phys. Rev. A*, 2003; 67 42105. DOI:10.1103/PhysRevA.67.042105.
[21] Erhart J., Sponar S., Sulyok G., Badurec G., Ozawa M. and Hasegawa Y. 'Experimental demonstration of a universally valid error–disturbance uncertainty relation in spin measurements'. *Nature Physics*, 2012; 8 185–189. DOI:10.1038/nphys2194.
[22] Gerlach W. and Stern O. 'Der Experimentelle Nachweis der Richtungsquantelung im Magneticfield'. *Z. Phys.*, 1922; 9 349–352. DOI:10.1007/BF01326983.
[23] Pauli W. 'Zur Quantenmechanik des magnetischen Elektrons'. *Z. Phys.* 1927; 43 601–623. DOI:10.1007/BF01397326.
[24] Klein O. 'Quantentheorie und fünfdimensionale Relativitätstheorie'. *Z. Physik*, 1926; 37 895–906. DOI:10.1007/BF01397481.
[25] Pauli W. and Weisskopf V. 'On quantization of the scalar relativistic wave equation'. *Helv. Phys. Acta.* 1934; 7 709–731.
[26] Dirac P. A. M. 'The quantum theory of electron'. *Proc. R. Soc. A*, 1928; 117 (778) 610–624. DOI:10.1098/rspa.1928.0023.
[27] Kusch, R. and Foley,H. M. 'The magnetic moment of electron'. *Phys. Rev.*, 1948; 74 250–263. DOI:10.1103/PhysRev.74.250.
[28] Schwinger J. 'On quantum-electrodynamics and the magnetic moment of the electron'. *Phys. Rev.*, 1948; 73 416–417. DOI:10.1103/PhysRev.73.416.
[29] Lamb W. E. and Retherford R. C. 'Fine structure of the hydrogen atom by a microwave method'. *Phys. Rev.*, 1947; 72 241–243. DOI:10.1103/PhysRev.72.241.
[30] Rarita W. and Schwiner J. 'On a theory of particles with half-integral spin'. *Phys. Rev.* 1941; 60 61. DOI:10.1103/PhysRev.60.61.
[31] Bargmann V. and Wigner E. P. 'Group theoretical discussion of relativistic wave equations'. *Proceedings of the National Academy of Sciences of the United States of America*; 1948, 34 211–223. DOI:10.1073/pnas.34.5.211.

Chapter 4
Atomic clock and lasers

Abstract

The measurement uncertainty of time and frequency has been drastically reduced since the development of atomic clocks. The first atomic clock was developed using the transition frequency in the microwave region because a narrow frequency bandwidth was obtained only in the microwave region in the 1950s. The uncertainty of 10^{-13} has been obtained with a commercial Cs atomic clock. The development of lasers made a great contribution to experimental physics. Atomic clocks using optical transition frequencies were realized after the development of laser spectroscopy, laser cooling, and frequency measurement of laser light. The uncertainty of 10^{-18} was obtained for several atomic transition frequencies. Lower measurement uncertainties are expected with the transition frequencies of highly charged ions or nuclei. An ultra-short pulse laser was also developed, which is useful to observe fast phenomena.

4.1 Atomic clocks in the microwave domain (pre-laser era)

As shown in Chapter 3.8, it has been clarified that the atoms and molecules absorb or emit electromagnetic waves (light, microwave) only when their frequency ν corresponds to the transition frequencies ν_0 satisfying

$$\nu_0 = \frac{\Delta E}{h} \tag{4.1}$$

where ΔE is the energy gap between different quantum states. The transition frequencies can be measured with uncertainty much lower than any other physical values as shown in Chapter 2.3, because the ultra-low statistical uncertainty is given by the ratio of transition frequency/spectrum linewidth, higher than 10^7 (Chapter 2.2.1). The atomic transition frequencies are determined by the Coulomb force between electron and nuclei (bonding force), which is uniquely determined. Electromagnetic perturbation gives a nonzero systematic uncertainty (Chapter 2.2.2), but it is much lower than the bonding force and the systematic uncertainty is also low. Therefore, atomic transition frequencies seemed to be a good standard for

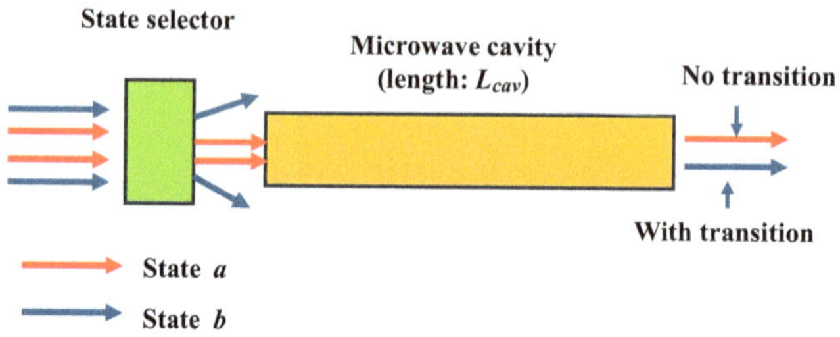

Figure 4.1 *Fundamental structure of Cs atomic clock using Cs atomic beam having states a and b. As the natural population in both states is almost equal (small energy gap), we need some procedure to localize atoms in one state. An atomic beam localized in a state passes through a microwave cavity. The transition rate is monitored from the population in state b at the exit.*

time and frequency. Atomic clocks were first developed using atomic transition frequencies in the microwave region because microwaves with narrow bandwidth were obtained in the 1950s. The microwave frequency could be measured with counters. There was no light source with a narrow frequency bandwidth until the laser was invented (Chapter 4.2).

In 1955, an atomic clock based on the transition frequency between hyperfine states (a and b) of the Cesium (Cs) atom was constructed. In state a (b), electron spin and nuclear spin are parallel (antiparallel). Why was the Cs atomic transition chosen as the standard? The Cs atom is a stable alkali atom (simple energy structure), with which the hyperfine transition frequency in the $^2S_{1/2}$ state is highest, and hence the time unit of 1 second can have a finer scale division. Therefore, the atomic transition frequency of Cs was best suited to achieve the highest accuracy.

The energy splitting between a and b is much smaller than $k_B T$ with a temperature of 300 K, and the populations in both states are almost equal (Chapter 3.4). The structure of Cs atomic clock is shown in Figure 4.1. Here, k_B is the Boltzmann constant and T is the thermodynamic temperature. To observe the transition between both states from the change of population, we need to localize atoms in one state as the initial state. An atomic beam flowing from an oven passes through a magnetic selector, with which atoms in a state are focused while atoms in b state are deflected. Atoms localized in a state pass through a microwave cavity with a length of L_{cav}. If the microwave frequency is resonant to the $a \rightarrow b$ transition, atoms in the b state are detected. The microwave frequency is stabilized to the transition frequency, giving feedback to maximize the $a \rightarrow b$ transition rate. The frequency uncertainty of 10^{-13} (300,000 years to give an error of 1 s) was obtained with the commercial type of Cs clock with $L_{cav} = 30$ cm. Lower systematic measurement uncertainty is obtained using laboratory-type clocks, giving corrections for the Zeeman shift induced by the magnetic field, relativistic effects, etc. (Chapters 2.2.2, 2.3, and 5.1). Lower statistic uncertainty (Chapter 2.2.1) is obtained by taking a larger L_{cav} (narrower linewidth). Measuring uncertainty of 1.2×10^{-14} has been obtained using an apparatus with $L_{cav} = 1.5$ m [1]. In 1967, the Cs transition frequency (9,192,631,770 Hz) was defined to be the standard of the time and frequency [2,3].

Except for the Cs clock, a hydrogen maser was developed, with which low statistical uncertainty is obtained with a short measurement time (frequency 1,420,405,751.7684 Hz) [4].

4.2 Lasers

A laser is a light source with the following characteristics

1. The phase is uniform, and it is useful to observe the interferometry.
2. The frequency bandwidth of each component is very narrow.
3. The propagation direction is also uniform.
4. The laser light can be focused to a spot size of the order of the wavelength.
5. Ultra-short pulses (10^{-15} s) can be obtained by the interference between many frequency components, which makes it possible to observe ultra-fast phenomena.

As shown above, a laser is a light source with which physical values are manipulated easily. The invention of the laser made a revolution in experimental physics, as shown below.

4.2.1 Principles of lasers

The word "LASER" is the abbreviation of "**L**ight **A**mplifier by **S**timulated **E**mission of **R**adiation." The transition between two energy states, 1 and 2, having energies of E_1 and E_2 ($E_1 < E_2$) and the populations $\rho(1)$ and $\rho(2)$. For simplicity, the degeneracies at both states are assumed to be 1. With a thermal equilibrium state, $\rho(1) > \rho(2)$ as shown in Chapter 3.4. By the irradiation of a light with the transition frequency $\nu_0 = [E_2 - E_1]/h$, the $1 \to 2$ transition is induced with the rate of $B_t I_{light}(\rho(1) - \rho(2))$, where B_t is the coefficient, and I_{light} is the intensity of the incident light wave. The incident light wave is damped with this transition so that the total energy is conserved as shown in Figure 4.2. There is a $2 \to 1$ transition emitting a fluorescent light with the rate of $A_t\rho(2)$, which is called a spontaneous emission transition. The phase and the direction of the emitted fluorescence light are totally random. The equilibrium population is given by $B_t I_{light}(\rho(1) - \rho(2)) = A_t\rho(2)$. When the transition is induced by the blackbody radiation (Chapter 3.5.2), $\rho(2)/\rho(1) = \exp[-(E_2 - E_1)/k_B T]$.

When $\rho(1) < \rho(2)$ (called inversion population), the $2 \to 1$ transition is more dominant than the $1 \to 2$ transition. The incident light wave is amplified with the same phase and the propagation direction as shown in Figure 4.2. Pumping to state 2 (some artificial treatment to increase $\rho(2)$) is required to obtain the inversion

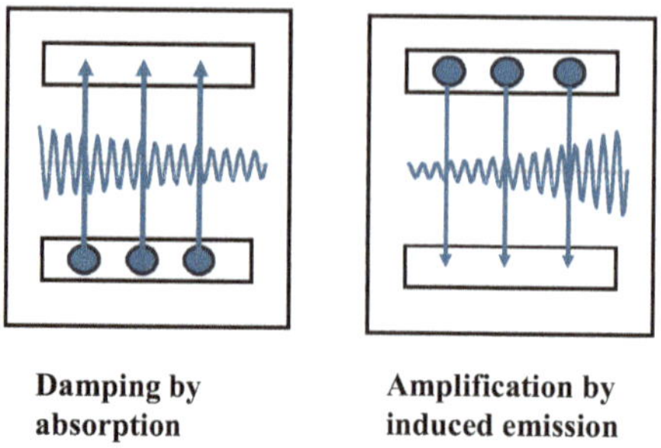

Figure 4.2 *The light with the transition frequency induces the transition between two energy states of the medium. When the medium is in the lower energy state, light is damped by absorption. When the medium is in the higher state, light is amplified by induced emission.*

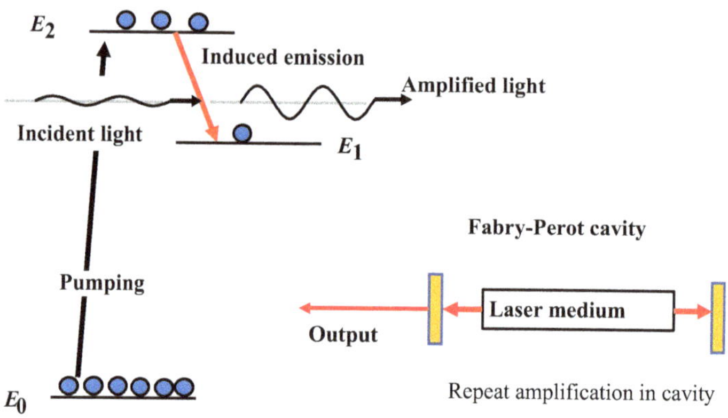

Figure 4.3 *Pumping with a three-state system. Pumping from the state 0, the population in the state 2 is higher than that in the state 1, and the incident light is amplified. When the laser medium is in a Fabry-Perot cavity, the amplification repeats in the cavity.*

population between the 1 and 2 states. It is realistic to attain the inversion population with a three-state system. We consider three energy states, 0, 1, and 2, with energies of E_0, E_1, and E_2 ($E_0 < E_1 < E_2$) and populations of $\rho(0)$, $\rho(1)$, and $\rho(2)$. When $\rho(0) \gg \rho(1)$, the inversion population between 1 and 2 states is obtained by the $0 \rightarrow 2$ pumping with the realistic rate as shown in Figure 4.3.

A laser is mostly used as a light source containing the laser medium in a cavity as shown in Figure 4.3 (Fabry-Perot cavity is considered in this figure). When the amplification (gain) of the light given by the laser medium is higher than the loss inside the cavity, the light with the resonant mode by the cavity repeats amplification while reflecting in the cavity. As the light intensity is increased, $\rho(2) - \rho(1)$

becomes smaller due to the higher stimulated emission rate (saturation effect), and the gain becomes smaller. The equilibrium light intensity is given by the balance between the gain with the saturation effect and the cavity loss. The wavelength of the light resonant to the Fabry-Perot cavity with the length of L_c is $2L_c/n_m$, where n_m is an integer; therefore, the resonance frequency is $\nu_c(n_m) = n_m c/2L_c$. The frequency interval between different frequency components is $\nu_{rep} = c/2L_c$, which indicates the repetition rate of reflection in the cavity. The linewidth of each frequency component is $(\Delta\nu_c) = c/2L_cN_r$ (N_r: mean reflection time), which converges to zero with $L_cN_r \rightarrow \infty$. Q value given by $\nu_c(n_m)/(\Delta\nu_c) = n_m N_r$, which represents the number of oscillations while the photon reflects in the cavity. When the laser oscillation is possible only with a single resonance frequency, the laser oscillates with a single frequency. To select a single resonance frequency, a grating mirror (high reflectivity only with a limited frequency area) is often used.

Frequency transformation of laser lights is performed to obtain light sources with frequencies that are difficult to obtain from direct laser oscillation. When laser light interacts with a crystal, integer multiple frequency components are obtained. Frequency-doubled laser light is particularly often used as the light source in the ultraviolet region. Irradiating two laser lights with the frequencies of $\nu_{1,2}$ to a crystal, the light with the frequency of $|\nu_1 \pm \nu_2|$ is obtained. The light source in the infrared or far-infrared region is often obtained from the frequency difference between the two laser lights. To give a frequency shift to the laser frequency, an acoustic-optical-modulator (AOM) and an electric-optical-modulator (EOM) are often used. AOM gives the shift of 10–100 MHz and EOM gives the shift of several GHz, respectively.

Laser light can be transformed to far-distant places (several thousand km) by propagation in an optical fiber, which is a thin conductor (core) coated by another material (clad). The refractive index of the clad is lower than that of the core, so the leak of light from the core is totally reflected at the boundary surface. Laser light with a wavelength of 1.4–1.6 μm is often used to transmit to a distant place via optical fiber because of low power loss by the fiber material.

4.2.2 Optical pumping

We can manipulate the quantum state of atoms or molecules using laser light. The population in the different quantum states in atoms or molecules is almost the same when the energy difference between them is smaller than k_BT as shown in Chapter 3.4 (k_B: Boltzmann constant, T: thermodynamic temperature). Optical pumping is a method to localize them in a single state. Here, we consider two ground states $g_{1,2}$ and one excited state e. When the $g_2 \rightarrow e$ transition is induced by a laser light, the deexcitation is caused by the spontaneous emission transitions to the g_1 or g_2 state. The $g_2 \rightarrow e \rightarrow g_1$ transition is caused by this procedure, while no transition is induced from the g_1 state. The selective excitation from the g_2 state is possible because of the frequency bandwidth of the laser light narrower than the transition frequency between g_1 and g_2. When there is no energy gap between g_1 and g_2, a selective transition can be performed by using a laser light with a

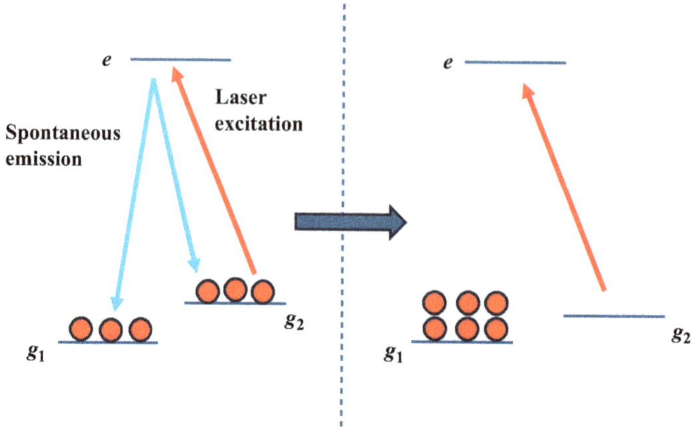

Figure 4.4 The procedure of optical pumping. Laser light is irradiated to induce the $g_2 \rightarrow e$ transition, while it does not induce $g_1 \rightarrow e$ transition. The spontaneous emission transition occurs in both g_1 and g_2 states. After repeating this cycle, all atoms are localized in g_1 state.

polarization that prohibits the transition from g_1. The excitation from the g_2 state repeats until all atoms are transformed to the g_1 state, as shown in Figure 4.4. No fluorescence is observed after the optical pumping. Optical pumping has been used, for example, with Cs atomic clock to localize atoms in one of two hyperfine states before interacting with microwaves [5].

Optical pumping is a technology to reduce the entropy (parameter of randomness) of atoms or molecules. However, the increase in entropy is obtained by transforming the laser light with a uniform phase to a fluorescence light with a random phase. Therefore, the principle of "increase of entropy (Chapter 3.4)" is not violated.

4.2.3 Laser spectroscopy

Measurements of the transition frequencies of atoms or molecules are useful to search their quantum energy structures, from which we can obtain information about constructing particles. The detailed information on transition frequencies of atoms and molecules also gives useful information for searching for the structures of stars. The frequency bandwidth of laser light is narrow, and transition frequencies are measured with high accuracy. The signal/noise ratio is increased drastically by using lasers; therefore, transitions with low rates can also be observed. The control of the optical path is easy when laser light is used because of its uniform propagation direction. The optical alignment becomes simpler using an optical fiber.

Since 1970, the spectroscopic research of polar molecules (NH_3 [6], H_2CO [7], CH_3F [8], etc.) has been performed, aimed at the analysis of their quantum energy structures. Until the development of frequency-tunable lasers, CO_2 or N_2O lasers

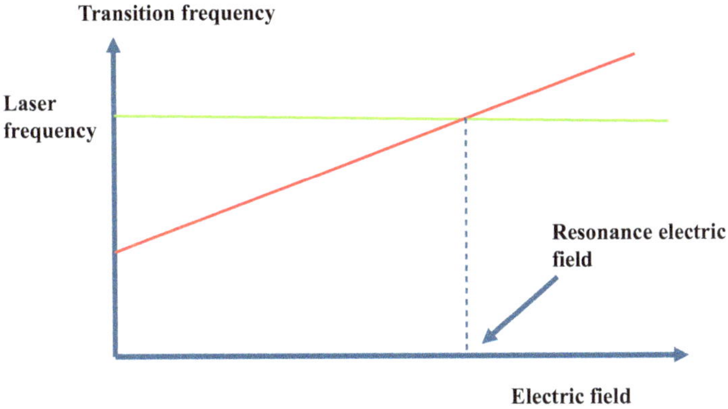

Figure 4.5 *Laser Stark spectroscopy using CO_2 or N_2O lasers (discrete frequencies), giving the shift of the transition frequency by a DC electric field*

with many discrete oscillation frequencies were used for experiments. The spectra were observed by inducing the shift of the transition frequencies by applying a DC electric field (laser Stark spectroscopy). The quantum energy structure was analyzed from the combination of laser frequencies and electric field strength, where the transition is induced as shown in Figure 4.5. Frequency-tunable dye lasers and diode lasers were developed afterwards, and the observation of the spectrum by frequency scanning became possible [9].

Atomic or molecular spectra are also used for the stabilization of laser frequency by giving feedback to maximize the transition rate, which also gave a chance to develop atomic clocks based on the transition frequencies in the optical region. However, it was not realistic to be applied for atomic clock, while the laser frequency was measured using a wavelength meter with an uncertainty of 10^{-7}. Laser spectroscopy was applied to atomic clocks after the development of the frequency comb (Chapter 4.2.5).

4.2.4 Laser cooling

The development of laser cooling technology made it possible to decelerate gaseous atoms from the speed of 200–400 m/s to slower than 10 cm/s. Here, we assume an atom with one ground state and one excited state. When atoms absorb laser light (energy and momentum of photons), they are transformed into an excited energy state and experience a force along the direction of light propagation, as shown in Figure 4.6 [10]. Afterwards, atoms emit photons in random directions (spontaneous emission) and return to the ground state. Atoms receive a recoil force in the inverse direction to the emitted photons, which have an average of zero. Repeating the cycle of absorption and spontaneous emission, atoms experience a net force along the direction (called "scattering force") of light propagation. The

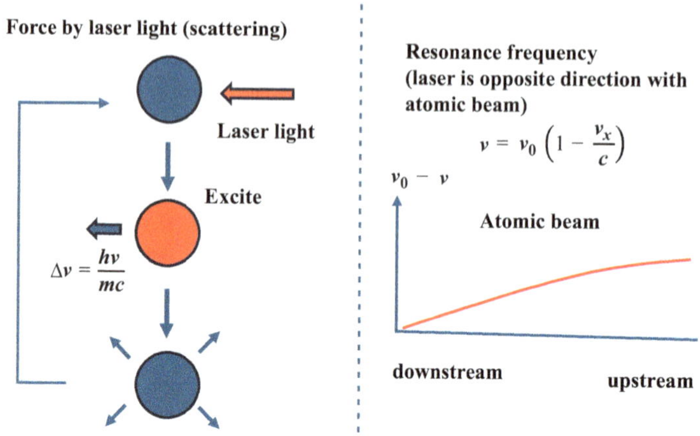

Figure 4.6 *The scattering force given by the cycle of laser excitation and spontaneous emission transition. At the laser excitation, atoms absorb the momentum of the photon. The force at the spontaneous emission transition is, on average, zero, and atoms get force in the propagation direction of laser light. Atomic beams can be decelerated by irradiating a laser beam in the opposite direction. To get a high deceleration effect, $\nu_0 - \nu$ (ν_0: transition frequency, ν:laser frequency) should be taken large at upstream (interact with atoms with high velocity) and decreased at downstream so that it works for slow atoms.*

force along the direction (here we consider the x-direction) of the laser light is as follows:

$$F_x = \frac{h\nu}{c} R(\nu)$$

$$R(\nu) \propto \frac{\Gamma}{\Delta^2 + \Gamma^2} \qquad \Delta = \nu_0 - \nu$$

ν: laser light frequency $\quad \nu_0$: transition frequency

Γ: spectrum linewidth given by spontaneous emission rate ($2\pi\Gamma$)

$R(\nu)$: rate of the cycle of excitation and spontaneous emission (4.2)

Therefore, atoms in a beam with a uniform direction can be decelerated using one laser in the opposite direction. Considering the Doppler effect, the deceleration is significant when the laser frequency ν is $\nu_0[1 - (v_x/c)]$, where v_x is the atomic velocity component parallel to the laser beam. The optimum value of $\Delta(= \nu_0 - \nu)$ is large for the initial high velocity and it becomes smaller as the atoms are decelerated as shown in Figure 4.6. To decelerate atoms from several hundred m/s to below 1 m/s with high efficiency, two methods are mainly used. One method is the application of an inhomogeneous magnetic field to give a shift of the transition

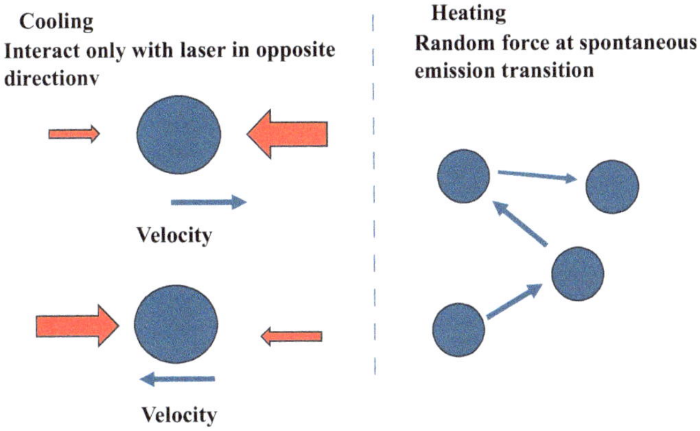

Figure 4.7 *Cooling by two laser lights with red detuning ($\nu_0 - \nu > 0$) from opposite directions. Atoms with any velocity interact with lasers with a direction opposite to the motion direction and are decelerated. Atoms are cooled in this procedure. There is also a heating effect caused by the random force at the spontaneous emission transition.*

frequency so that Δ is large at upstream and small at downstream. This system is called "Zeeman slower." Another method is frequency scanning of deceleration laser. The average velocity of the atoms in the beam can be reduced to close to zero with these methods. But there are still velocity components in random directions. The laser light accelerates atoms with a motion parallel to the laser light direction. The next step is required for further cooling.

To cool atoms with random motion direction, two laser beams with $\Delta > 0$ (red detuned) irradiate the atoms from opposing directions. Because of the Doppler effect, atoms get a stronger force from the laser light in the opposite direction to the motion and all atoms are decelerated as shown in Figure 4.7. The change of kinetic energy of an atom E_K with the velocity of $v_x = \pm\sqrt{2E_K/m}$ is given by

$$\frac{dE_K}{dt} = mv_x\frac{dv_x}{dt} = v_xF_x - \frac{h\nu}{c}v_x\left[R\left(\nu\left(1+\frac{v_x}{c}\right)\right) - R\left(\nu\left(1-\frac{v_x}{c}\right)\right)\right]$$

$$R\left(\nu\left(1\pm\frac{v_x}{c}\right)\right) \approx R(\nu)\left(1\mp\frac{2\nu(v_x/c)\Delta}{\Delta^2+\Gamma^2}\right)$$

$$\frac{dE_K}{dt} = -R(\nu)\frac{h\nu^2}{mc^2}\frac{8E_K}{\Delta^2+\Gamma^2}\Delta. \tag{4.3}$$

Equation (4.3) indicates that the kinetic energy reduces when $\Delta > 0$.

As the atoms decelerate and the Doppler shift becomes smaller than Γ, the difference in the interaction forces from the two opposing laser beams becomes smaller. There is also a heating effect caused by the momentum of photons in a random direction at spontaneous emission. Considering the random change of the momentum at spontaneous emission (Figure 4.7), there is a heating effect whose rate is given by (note that the transition cycle is caused by two lasers)

$$E_{scat} = \frac{1}{2m}\left(\frac{h\nu}{c}\right)^2 R(\nu) \times 2 = \frac{1}{m}\left(\frac{h\nu}{c}\right)^2 R(\nu). \tag{4.4}$$

Considering (4.3) and (4.4), the total change of the kinetic energy is given by

$$\begin{aligned}\frac{dE_K}{dt} &= -R(\nu)\frac{h\nu^2}{mc^2}\frac{8E_V}{\Delta^2+\Gamma^2}\Delta + \frac{1}{m}\left(\frac{h\nu}{c}\right)^2 R(\nu)\\ &= R(\nu)\frac{1}{m}\left(\frac{h\nu}{c}\right)^2\left[-\frac{8E_V}{h\{\Delta^2+\Gamma^2\}}\Delta + 1\right],\end{aligned} \tag{4.5}$$

and the equilibrium kinetic energy is given by

$$E_{V-eq} = \frac{h}{8\Delta}\left(\Delta^2+\Gamma^2\right), \tag{4.6}$$

which is minimum with $\Delta = \Gamma$. The minimum equilibrium energy (called the Doppler limit)

$$E_{Doppler} = \frac{h\Gamma}{4}. \tag{4.7}$$

Using the transition with a large Γ (high transition rate), the cooling force is strong because of the rapid repetition of the cooling transition cycle. Using transitions with small Γ (low transition rate), lower kinetic energy is attained. Two steps of cooling are sometimes performed: the first cooling state using a transition with high Γ (several MHz) for rapid cooling and the second cooling stage using a transition with low Γ (several Hz) to attain lower kinetic energy. When one-dimensional cooling is performed, heating with the motion in other directions is caused by the random force due to the spontaneous emission. For the two- or three-dimensional cooling, four or six cooling lasers are required.

Laser-cooled atoms fall under the gravitational force unless they are trapped in an area with some method. The expansion is also significant without the trapping force. The MOT is an apparatus to trap laser-cooled atoms in a small area using the scattering force of the cooling laser, as shown in Figure 4.8 [11]. Using an anti-Helmholz coil (a pair of coils giving magnetic fields in counterpropagating directions), a magnetic field proportional to x is given (x: displacement from the center). The Zeeman shift (energy shift by magnetic field shown in Chapter 3.9) for the atoms in the M_F state is given with a formula $ha_Z M_F x$. Here, M_F is the component of total angular momentum (electron spin, electron orbital angular momentum, and nuclear spin) parallel to the magnetic field and a_Z is the coefficient of the Zeeman shift. Two cooling lasers are irradiated from opposing directions. The polarization of one laser is circularly polarized in the right-handed (σ^+) and the transition rate has a dependence on M_F state and ν: $R_+(M_F, \nu)$. The other laser

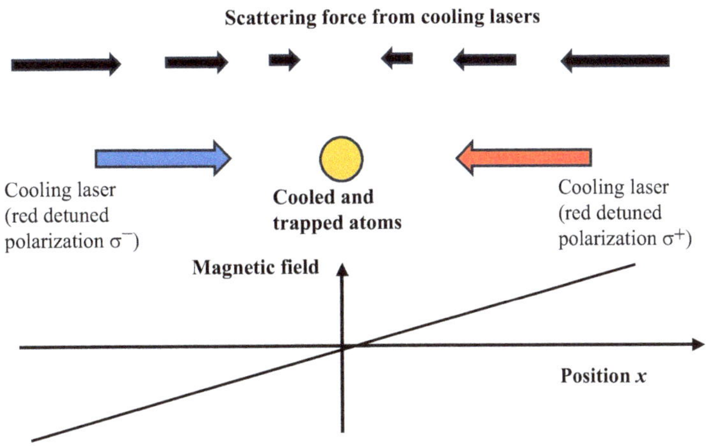

Figure 4.8 *The concept of a magnetic optical trap to give not only the cooling but also the trapping force*

is left-handed (σ^-) polarization and the transition rate is $R_-(M_F, \nu) = R_+(-M_F, \nu)$. Assuming that the intensities of two laser lights are equal, the populations in the $\pm M_F$ states are equal. The scattering force for the atoms in the $\pm M_F$ is given by

force from σ^+ laser

$$F_{x+} = \frac{h\nu}{c}\left[R_+(\nu, M_F)\left(1 - \frac{2\Delta\nu(v_x/c) + 2\Delta a_Z M_F x.}{\Delta^2 + \Gamma^2}\right)\right.$$

$$\left. + R_+(\nu, -M_F)\left(1 - \frac{2\Delta\nu(v_x/c) - 2\Delta a_Z M_F x.}{\Delta^2 + \Gamma^2}\right)\right]$$

force from σ^- laser

$$F_{x-} = \frac{h\nu}{c}\left[R_-(\nu, M_F)\left(1 - \frac{-2\Delta\nu(v_x/c) + 2\Delta a_Z M_F x.}{\Delta^2 + \Gamma^2}\right)\right.$$

$$\left. + R_-(\nu, -M_F)\left(1 - \frac{-2\Delta\nu(v_x/c) - 2\Delta a_Z M_F x.}{\Delta^2 + \Gamma^2}\right)\right]$$

$$= \frac{h\nu}{c}\left[R_+(\nu, -M_F)\left(1 - \frac{-2\Delta\nu(v_x/c) + 2\Delta a_Z M_F x.}{\Delta^2 + \Gamma^2}\right)\right.$$

$$\left. + R_+(\nu, M_F)\left(1 - \frac{-2\Delta\nu(v_x/c) - 2\Delta a_Z M_F x.}{\Delta^2 + \Gamma^2}\right)\right]$$

total force

$$F_{x+} - F_{x-} = -\frac{4hv^2}{c^2(\Delta^2+\Gamma^2)}\Delta[R_+(v,M_F)+R_+(v,-M_F)]v_x - \frac{4hva_ZM_F}{c(\Delta^2+\Gamma^2)}\Delta[R_+(v,M_F)-R_+(v,-M_F)]x. \tag{4.8}$$

Equation (4.8) indicates that the scattering force induces the damping oscillation at around $x = 0$ when $\Delta > 0$. Using lasers in six directions, MOT gives a three-dimensional cooling and trapping force.

Laser cooling is also applicable to ions, which are confined in a small area by an inhomogeneous RF electric field (Chapter 4.3.1). Trapped ions exhibit a vibrational motion around the center position with zero electric field (Chapter 4.3.1), and laser cooling is useful to reduce their amplitude, as shown in Figure 4.9. Laser cooling can be performed with one cooling laser with a frequency lower than the transition frequency [12]. Trapped ions repeat the motion parallel and opposite directions with the laser light. The interaction between the ion and laser light is significant only at the motion in the opposite direction because of the Doppler effect, and the amplitude of the vibrational motion is reduced. The motions of trapped ions in three directions are coupled and cooled in all directions by a cooling laser in one direction.

We assumed that the spontaneous emission transition from the excited state is possible only to the initial ground state. When a spontaneous emission transition to another state is caused, the cooling cycle (laser excitation+ spontaneous emission) stops afterwards. A repump laser, giving the transition to the excited state, is needed to push atoms back to the cooling cycle. With complex energy states, more

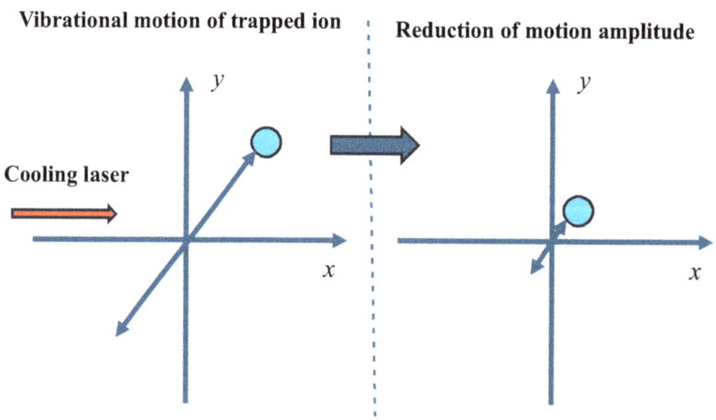

Figure 4.9 Ions trapped by an inhomogeneous AC electric field have vibrational motion around the center, where the electric field is zero. The amplitude of vibrational motion is reduced by laser cooling using a laser light in one direction.

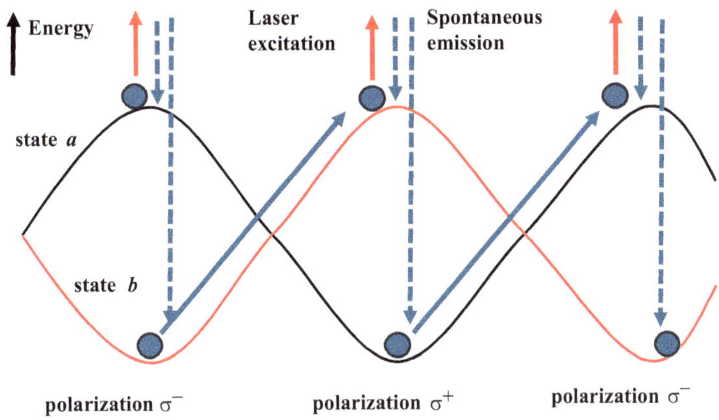

Figure 4.10 *The mechanism of polarization gradient cooling. The interaction with the cooling laser induces the energy splitting of the degenerated substates, depending on the polarization. The transition from the higher energy a, b state to the lower sub-state makes the energy loss of atoms. This transition repeats due to the periodic change of the polarization of the laser light.*

repump lasers are required to maintain the laser cooling process. Laser cooling has been performed mainly for atoms with simple energy structures, such as alkali or alkali-earth atoms.

The kinetic energy below the Doppler limit was attained with atoms with which there is degeneracy in the ground state (multi-energy states with the same energy), which was a mystery until the mechanism of polarization gradient cooling was clarified [13]. Here, we consider substates a and b. The electric field of the cooling laser induces different energy shifts in the a and b state, which depend on the polarization (direction of the electric or magnetic field) of the laser light, as shown in Figure 4.10. The two cooling lasers with opposite directions form a standing wave and two circular polarizations ($\sigma^{\pm}$) appear periodically. With one polarization, the energy in the sub-state a is higher than b. Via the absorption transition from a to the excited state and the spontaneous emission transition to b, there is a loss of atomic potential energy because of the energy gap between a and b states. The excitation from b is not induced with this polarization. Atoms in b state move to the area of another polarization where the potential energy in b state is higher than in a state. The atomic kinetic energy is transformed into potential energy in this procedure. Then the transition from b to a is caused and the atomic potential energy is lost again. The cooling continues until atomic kinetic energy becomes lower than the energy splitting between the two substates. The attainable kinetic energy is roughly estimated by the one-photon recoil energy, $(h\nu_0)^2/2mc^2$ and the kinetic temperature of the order of a few μK is obtained with Cs and Rb atoms.

For atoms or ions trapped by a force giving a harmonic motion with the frequency of ν_{vib} the motion energy E_{vib} is given by (Chapter 3.8.4)

$$E_{vib}(n_{vib}) = \left(n_{vib} + \frac{1}{2}\right) h\nu_{vib} \quad n_{vib}\text{: integer.} \tag{4.9}$$

The transition is induced not only when the laser frequency corresponds to the transition frequency ν_0, but also with the sidebands given by

$$\nu_S(\Delta n_{vib}) = \nu_0 + \Delta n_{vib} \nu_{vib}, \tag{4.10}$$

because of the first-order Doppler effect. Each sideband component has the broadening of Γ. When the atomic transition is induced by the laser light with $\nu_S(\Delta n_{vib})$, the motion energy changes from $E_{vib}(n_{vib})$ to $E_{vib}(n_{vib} + \Delta n_{vib})$. There is a method called "sideband cooling" to reduce the motion energy by $h\nu_{vib}$ per cycle by irradiating the laser light with the frequency of $\nu_S(-1)$, as shown in Figure 4.11 [14]. The transition rates induced by the sideband $\nu_S(-1)$ and $\nu_S(1)$ are proportional to n_{vib} and $(n_{vib} + 1)$, respectively. With this method, the motion energy is reduced to the ground state $h\nu_{vib}/2$ (called zero-point energy), from where the transition rate induced by the sideband $\nu_S(-1)$ is zero. Sideband cooling is possible for transitions whose linewidth Γ is narrower than ν_{vib}, so that the neighboring sidebands are not overlapped. Figure 4.11 also indicates sideband Raman cooling [15] with the procedure of the Raman transition of $(g_1, n_{vib}) \to (g_2, n_{vib} - 1)$ and the optical pumping $(g_2, n_{vib} - 1) \to (g_1, n_{vib} - 1)$. Repeating this cycle, the energy of trapped atoms or ions can be reduced to the zero-point energy.

There is also a heating effect by the photon recoil energy $(h\nu_0)^2/2mc^2$. To attain the kinetic energy below the photon recoil energy, it is preferable to suppress the interaction with laser light for the atoms with the lowest kinetic energy. One method is Raman cooling using lasers A (frequency ν_a) and B (frequency ν_b) in the

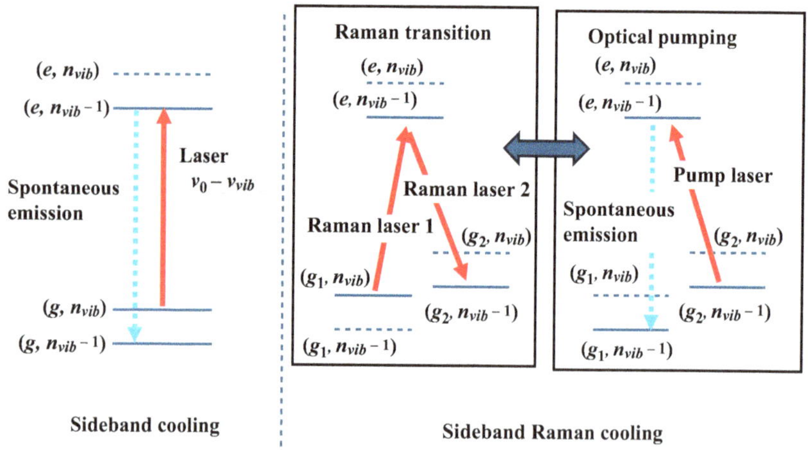

Figure 4.11 The transition procedure of sideband cooling and sideband Raman cooling

counterpropagating directions. A hyperfine transition (transition frequency ν_m) is induced when the atomic velocity v_x satisfies

$$|\nu_a - \nu_b| = \nu_m + \frac{v_x(\nu_a + \nu_b)}{c}. \tag{4.11}$$

With the one Raman transition, atoms get the momentum of $h(\nu_a + \nu_b)/c$. To repeat the cooling transition, atoms are pushed back to the initial state by optical pumping. This transition is induced only with atoms in a very narrow velocity region, because it is very sensitive to the first-order Doppler effect. Mean kinetic energy below one-photon recoil energy can be attained because atoms with zero velocity cannot interact with laser light [16].

4.2.5 Frequency combs

For the frequency precision measurement, optical transition frequencies are more advantageous than those in the microwave region because the time of 1 s is divided on a finer scale. However, atomic clocks using optical transition frequencies were not realized soon after the development of laser spectroscopy because the frequency of laser light was measured using a wavelength meter with an uncertainty of 10^{-7}.

We can measure the microwave frequency using a counter. Measurement of optical frequency is also possible as an integer multiple of a frequency in the microwave region.

The development of frequency combs made it possible to measure the frequency of laser light with an uncertainty below 10^{-18}. Chapter 4.2.1 indicates that the resonance frequency of a cavity (length L_c) is given by an integer multiple of $\nu_{rep}(= c/2L_c)$. When the laser oscillation is possible for the frequency components of $N_0 - \Delta N_m \leq n_m \leq N_0 + \Delta N_m$ with a uniform phase, the laser light is a pulse laser with a repetition rate of ν_{rep} and the length of $1/(\Delta N_m)\nu_{rep}$. The actual frequency components are given as

$$\nu_{com}(n_m) = n_m \nu_{rep} + \nu_{ceo}, \tag{4.12}$$

because the refractive index of the laser medium has a slight dependence on the frequency and a nonzero offset frequency ν_{ceo} exists. Using a frequency counter, ν_{rep} can be measured as the repetition rate of the pulsed laser. Also, ν_{ceo} can be measured from the beat signal between $2\nu_{com}(n_m)$ (frequency doubling shown in Chapter 4.2.1) and $\nu_{com}(2n_m)$.

The frequency of some laser light ν_L is measured from the minimum beat frequency $\nu_{beat} = |\nu_{con}(n_m) - \nu_L|$ after determining n_m using the wavelength meter as shown in Figure 4.12. Measurement with different ν_{rep} is useful to determine the detail of n_m and sign of $\nu_{com}(n_m) - \nu_L$. Development of optical atomic clocks became one of the hottest research themes since the development of the frequency comb. The frequency components can be stabilized by locking ν_{rep} and ν_{ceo} using an atomic clock in the microwave region. However, it is more common to lock one frequency component using an optical atomic clock, and then all frequency

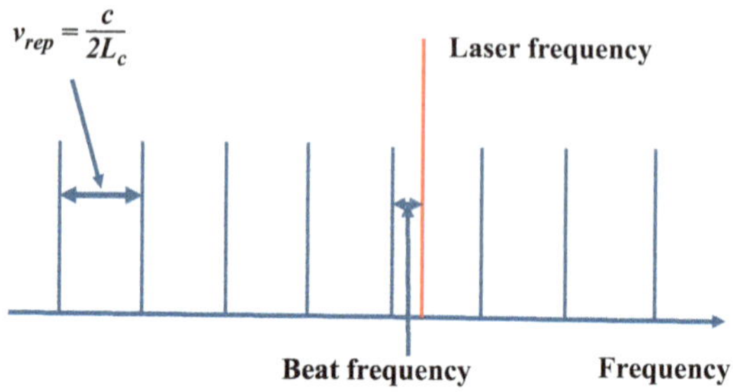

Figure 4.12 *Frequency components in a frequency comb are distributed with the interval of the repetition rate of the pulse laser $\nu_{rep} = c/2L_c$. The frequency of a laser light is measured from the beat frequency with the closest frequency component.*

components are locked simultaneously. Measuring other optical transition frequencies using this frequency comb, the ratio between two optical transition frequencies is determined with the uncertainties of 10^{-18}.

4.3 Optical atomic clocks

Suppression of the first-order Doppler effect is required for the development of atomic clocks. For the use of transition frequencies in the optical region, atoms are required to be confined in an area smaller than the wavelength of the probe laser. Atoms in a trapping field have vibrational motion, whose amplitude is reduced to below the optical wavelength by laser cooling. Quadratic Doppler effect (Chapters 2.3 and 5.1) is also reduced by using cold atoms.

The trap force induces the shift in the transition frequency if the trapping potential energies at the upper and lower states are different. Two methods were developed to give a trap force, with which the trapping potential energies at upper and lower levels are equal and this effect is eliminated for the measurement of the transition frequency.

The spectrum linewidth of atomic transitions in the optical region is mainly determined by the spontaneous emission rate because the random phase jump is caused by the spontaneous emission. Therefore, transitions with a high spontaneous emission rate (several MHz) are not advantageous for precision measurement because of high statistical uncertainty. This chapter introduces the precision measurements of transition frequencies whose spontaneous emission rates are so low (< 10 Hz) that the ratio transition frequency/line width is larger than 10^{13}.

4.3.1 Ion traps with RF electric fields

One method is to use ions, which are three-dimensionally trapped using an inhomogeneous RF electric field. The ion is trapped by the Coulomb force, which does not give any shift in the transition frequency (the Coulomb force is exactly the same for any quantum states).

Here, we introduce the mechanism by which ions can be trapped by an inhomogeneous RF electric field tightly around a position where the electric field is zero. Trapping and expanding forces are periodically applied by the RF electric field $\overrightarrow{E_e}(\vec{r})\sin(\Omega t)$. The motion equation of a charged particle is given by

$$m\frac{d^2\vec{r}}{dt^2} = q_e\overrightarrow{E_e}(\vec{r})\sin(\Omega t). \tag{4.13}$$

When the change in position within the period of the RF electric field $(\delta\vec{r})$ is negligibly small compared to $|\vec{r}|$ ($\overrightarrow{E_e}(\vec{r_0} \pm \delta\vec{r}) \approx \overrightarrow{E_e}(\vec{r_0})$), the motion of the ion is obtained as follows:

$$\vec{r} = \vec{r_0} + \delta\vec{r}$$

$$m\frac{d^2\delta\vec{r}}{dt^2} = q_e\overrightarrow{E_e}(\vec{r_0})\sin(\Omega t)$$

$$\delta\vec{r} = -\frac{q_e\overrightarrow{E_e}(\vec{r_0})}{m\Omega^2}\sin(\Omega t) + c_1 t + c_0$$

($c_{0,1}$: constant given by the initial position and velocity)

$$\left[m\frac{d^2\vec{r_0}}{dt^2}\right]_{ave} = \left[q_e\overrightarrow{E_e}(\vec{r_0} + \delta\vec{r})\sin(\Omega t)\right]_{ave}$$

$$= \left[q_e\overrightarrow{E_e}(\vec{r_0})\sin(\Omega t)\right]_{ave} + \left[q_e\frac{d\overrightarrow{E_e}(\vec{r_0})}{d\vec{r}}\delta\vec{r}\sin(\Omega t)\right]_{ave}$$

$$= -\left[q_e\frac{d\overrightarrow{E_e}(\vec{r_0})}{d\vec{r}}\left\{\frac{q_e\overrightarrow{E_e}(\vec{r_0})}{m\Omega^2}\sin(\Omega t)^2 + (c_1 t + c_0)\sin(\Omega t)\right\}\right]_{ave}$$

$$= -\frac{q_e^2\overrightarrow{E_e}(\vec{r_0})}{2m\Omega^2}\frac{d\overrightarrow{E_e}(\vec{r_0})}{d\vec{r}}. \tag{4.14}$$

The time-averaged force is nonzero because $\left|\overrightarrow{E_e}\left(\vec{r_0} + \delta\vec{r}\right)\right| \neq \left|\overrightarrow{E_e}\left(\vec{r_0} - \delta\vec{r}\right)\right|$, as shown in Figure 4.13. The motion of the ion (change of r_0) given by (4.14) is interpreted as the motion in the pseudopotential given by

$$P_{ps}(\vec{r_0}) = \frac{\left|q_e\overrightarrow{E_e}(\vec{r_0})\right|^2}{4m\Omega^2}. \tag{4.15}$$

$P_{ps}(\vec{r_0}) \geq 0$ and there is a force to confine the ion at the position where $\left|\overrightarrow{E_e}(\vec{r_0})\right|$ is minimum. Many apparatuses for ion trapping are designed so that ions

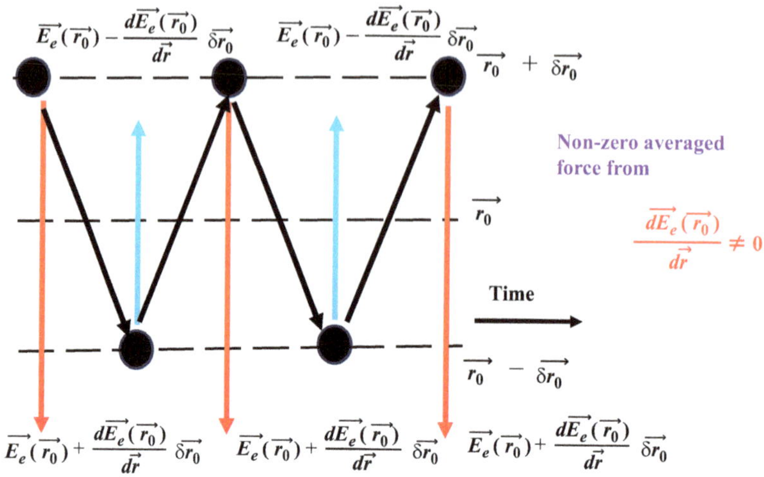

Figure 4.13 *Micromotion induced by AC electric field. Because the electric field is not homogeneous, the average of the force is not zero.*

are confined around $\vec{r_0} = 0$ where $\left|\vec{E_e}\right| = 0$. The amplitude of the micromotion $\overrightarrow{\delta r_0}$ is zero at $\vec{r_0} = 0$. Note that

$$\left|\vec{E_e}\left(\vec{r_0}\right)\right| \gg \left|\frac{d\vec{E_e}\left(\vec{r_0}\right)}{d\vec{r}}\delta\vec{r}\right| \rightarrow \frac{q_e}{m\Omega^2}\left|\frac{d\vec{E_e}\left(\vec{r_0}\right)}{d\vec{r}}\right| \ll 1 \tag{4.16}$$

is required for the use of (4.14).

For three-dimensional traps with an RF electric field, the distribution of $\vec{E_e} = C_{ac}\left(\alpha_x x, \alpha_y y, \alpha_z z\right)$ (Maxwell's equation requires $\alpha_x + \alpha_y + \alpha_z = 0$) is often used. There is also a linear trap apparatus giving an RF electric field of $\vec{E_e} = C_{ac}(x, -y, 0)$ in the x and y directions and a DC electric field of $\overrightarrow{E_{dc}} = G_{Dc}(0, 0, z)$ in the z-direction as shown in Figure 4.14. With the linear trap apparatus, the RF electric field is zero on the center axis.

The amplitudes of the vibrational motion of trapped ions are reduced to less than the optical wavelength by laser cooling. When a three-dimensional RF electric trapping system is used for precision measurement of transition frequency, the measurement should be performed with a single ion so that it is localized at the trap center $(x, y, z) = (0, 0, 0)$, where the RF electric field is zero. Using multi-ions, positions of trapped ions are shifted from the center by the Coulomb repulsion force between ions, and a significant shift in the transition frequency is induced by the RF electric field. Linear traps can be used for precision measurement of transition frequency using multi-ions because laser-cooled ions form a string crystal on the axis, where the RF electric field is zero. The DC electric field in the z-direction is also zero because $\overrightarrow{E_{DC}}$ is balanced with the electric field from the neighboring ion. However, the DC electric quadrupole shift (interaction between the electric

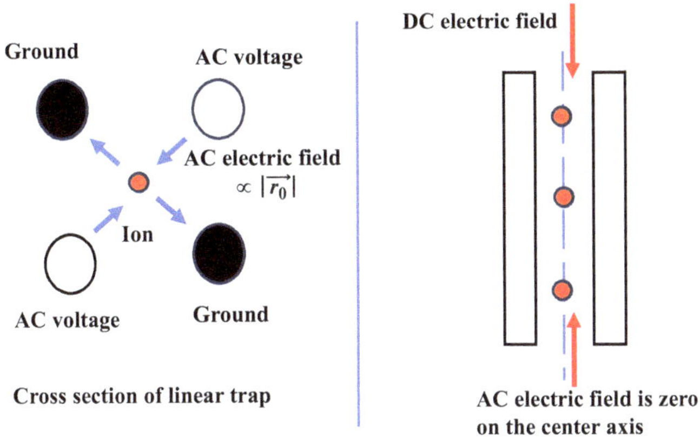

Figure 4.14 Structure of linear ion trap apparatus. A string crystal is formed by several ions that are trapped and laser-cooled.

quadrupole moment ξ_q and electric field gradient G_{DC}) is significant for ions with $\xi_q \neq 0$. Therefore, measurements using linear traps are applicable only for transition frequencies between states with $\xi_q = 0$.

The following transition frequencies have been measured with uncertainties of 10^{-18}.

Three-dimensional RF electric trapping

$^{171}Yb^+$ $^2S_{1/2}$ F = 0–$^2F_{7/2}$ F = 3 642, 121, 496, 772, 645.12 Hz 3×10^{-18} [17]

Linear trap

$^{115}In^+$ $^1S_0\ F = \frac{9}{2} - {}^3P_0\ F = \frac{9}{2}$ 1, 267, 402, 452, 901, 039.87 Hz 4.4×10^{-18} [18]

$^{27}Al^+$ $^1S_0\ F = \frac{5}{2} - {}^3P_0\ F = \frac{5}{2}$ 1, 121, 015, 393, 207, 859.16 Hz 5.5×10^{-19} [19]

4.3.2 Neutral atoms in optical lattices

A method was also developed to measure the $^1S_0 \rightarrow {}^3P_0$ transition frequencies of neutral alkali-earth atoms. The atoms are laser cooled by two steps: after rapid cooling using the $^1S_0 \rightarrow {}^1P_1$ transition, second-stage cooling using the $^1S_0 \rightarrow {}^3P_1$ transition attains the kinetic energy of 1 μ K. Atoms in the magnetic optical trap (MOT: Chapter 4.2.4) are transformed to the trapping at the antinodes of the standing wave of the laser light (an optical lattice) as shown in Figure 4.15. The trap potential is given by the interaction with the electric field of the trap laser light (strong at antinodes and weak at nodes). The shift in the transition frequency is given by the difference between the trap potential energies in the upper and lower states, which can be positive or negative depending on the trap laser frequency.

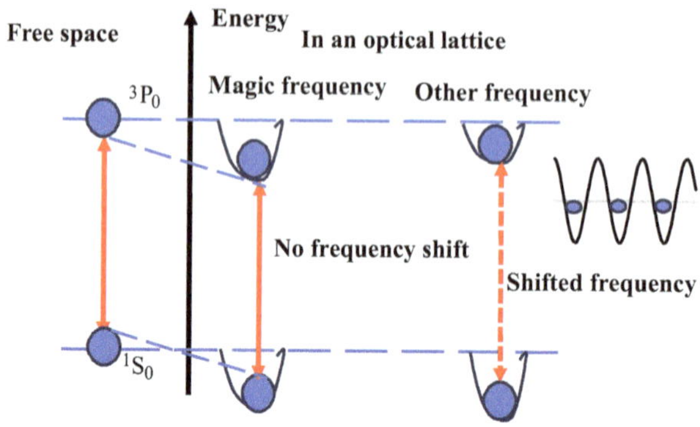

Figure 4.15 The fundamentals of the optical lattice clock. Trap force is given by the energy shift induced by the electric field of the trap laser. Choosing the trap laser frequency (magic frequency), the energy shifts at upper and lower states are equal, and the shift in transition frequency is eliminated.

Choosing the trap laser frequency where the potential energies in both states are equal, the shift in the transition frequency is eliminated (called the magic frequency) as shown in Figure 4.15. When the trap laser frequency is shifted from the magic frequency, a shift in the transition frequency exists and it is proportional to the trap laser intensity. For the $^1S_0 \rightarrow ^3P_0$ transition frequencies of alkali-earth atoms, the fractional shift in the transition frequency is less than 10^{-18} when the trap laser frequency is shifted from the magic frequency by 1 MHz and the trap laser intensity is the value to make the potential depth of 10 μK. Measurements are performed using many atoms to achieve low statistical uncertainty with a short measurement time.

The measurement has been performed with $^1S_0 - ^3P_0$ transition frequencies of ^{87}Sr, ^{171}Yb, and ^{199}Hg atoms. The measured frequencies were [20]

^{87}Sr $429,228,066,418,007.01$ Hz

^{171}Yb $518,295,836,590,863.63$ Hz

^{199}Hg $1,128,575,290,808,154.32$ Hz.

The Boulder group measured the ratio between the $^1S_0 \rightarrow ^3P_0$ transition frequencies with the uncertainties as follows [21].

$^{27}Al^+ - {}^{87}Sr\ 8.0 \times 10^{-18}$

$^{27}Al^+ - {}^{171}Yb\ 5.9 \times 10^{-18}$

$^{27}Sr - {}^{171}Yb\ \ 6.8 \times 10^{-18}$

In principle, the uncertainty of the ^{199}Hg transition frequency can be further reduced. However, it took time until the measurement was performed, because the trap and probe lasers must operate in the ultraviolet region, which is obtained by frequency doubling of laser light in the optical region.

The recent measurement uncertainty of the ^{87}Sr transition uncertainty was reduced to 8.1×10^{-19} [22].

4.3.3 Future developments in atomic clocks

The atomic transition frequency is determined mainly by the bond force between the electron and the nucleus, which is uniquely determined. The existence of electromagnetic noise induces systematic uncertainty, and the transitions with a higher ratio of bonding force/electromagnetic noise are advantageous for precision measurement. For the $^1S_0 - {}^3P_0$ transitions, lower systematic uncertainties were previously obtained for single-charged ions ($^{115}In^+$, $^{27}Al^+$) than neutral atom (^{87}Sr, ^{171}Yb), because of the stronger Coulomb interaction between electron and nucleus. Lower statistical uncertainty was obtained for neutral atoms by measuring using many atoms in an optical lattice.

The transition frequencies of highly charged ions are expected to be measured with lower systematic uncertainties than previous atomic clocks because of the stronger bonding force between the electron and nuclear as shown in Figure 4.16. However, the measurements of these transition frequencies have been realized just recently because of technical difficulties.

Highly charged ions are produced when the initial kinetic energy is of the order of 10^6 K. Laser cooling of a highly charged ion is not realistic because there is no transition applicable for laser cooling in the optical region. Cooling of highly charged ions is performed using the Coulomb interaction with many single-charged ions, which were laser cooled (sympathetic cooling). The measurement results of

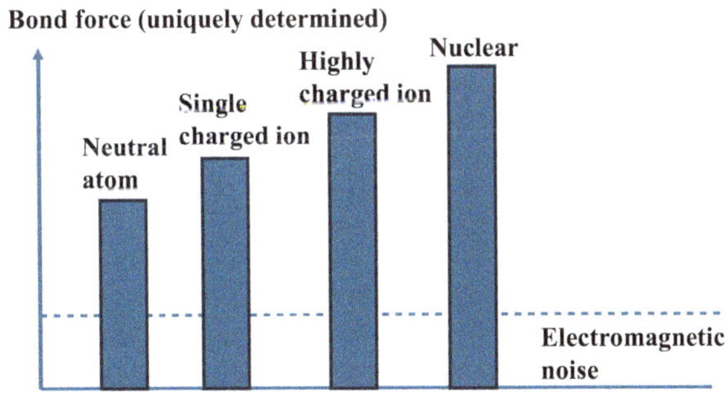

Figure 4.16 The transition frequencies are determined mainly by the bonding force, which is determined uniquely. Lower uncertainty is attainable with a larger ratio (bonding force/electromagnetic noise). Transition frequencies of highly charged ions and nuclei are more advantageous for precision measurement than previous atomic clocks.

the Ar^{13+} $^2P_{1/2} \rightarrow {}^2P_{3/2}$ transition frequency were [23]

$$^{40}Ar^{13+}\ 679,216,462,397,957.43\ \text{Hz, uncertainty} \qquad 1.5 \times 10^{-16}$$

$$^{36}Ar^{13+}\ 679,214,584,287,424.91\ \text{Hz, uncertainty} \qquad 1.7 \times 10^{-16}.$$

The systematic uncertainty is 2.2×10^{-17}, and the total uncertainty was dominated by statistical uncertainty, which can be reduced by taking a longer measurement time.

Precise measurements of nuclear transition frequencies are also of interest. The energy gap between different nuclear states is much larger than that for electron energy state because the protons and neutrons in the nucleus are bound in an area five orders smaller than the electron orbits. Most of the nuclear transition frequencies are in the γ-ray region, where frequency measurements are difficult. Nevertheless, the thorium-299 nucleus (^{299}Th) has a transition from the ground state to the first excited state in the vacuum ultraviolet (VUV) region, where the frequency can be measured using a frequency comb. The spontaneous emission rate is much lower than 1 Hz and the spectrum linewidth is limited just by that of the probe laser. Therefore, the statistical uncertainty is low because the ratio transition frequency/line width larger than 10^{15}. The systematic uncertainty is expected to be much smaller than atomic transition frequencies because the bonding force between protons and neutrons in the nucleus (nuclear force) is much stronger than the Coulomb force between electrons and nuclei. The clock transition frequency has been estimated by the frequency difference between γ-decay emitted from the nuclei. Until 2007, there were discrepancies between results reported by different groups. Recently, consistent results were reported by several groups. Elwell *et al.* reported the result of 2,020,407. 3 $\pm$3 GHz [24]. Zhang *et al.* performed a laser spectroscopic experiment and obtained the measurement result of 2,020,407,384,335$\pm$ 2 kHz [25].

4.3.4 *Attainable uncertainties in the future*

Chapter 4.3.3 indicates the possibility of suppressing the measurement uncertainties induced by electromagnetic noise. Then can we get the uncertainty below 10^{-20}? There is a serious problem to overcome, as shown below.

There is a frequency shift induced by the gravitational potential (Chapters 3.7 and 5.1), which gives the shift with the ratio of 1.1×10^{-16} with the change of altitude of 1 m. When the measurement uncertainty is reduced to 10^{-19}, the fluctuation due to the change of altitude of 1 mm is detected. The gravitational potential is not perfectly constant, even if we do not change the place of the atomic clock. For example, the influence of Earth's tide (change of gravitational potential from the Moon) has already been detected.

Previously, the frequency standard was defined as that on the geoid surface, but it cannot be defined with an uncertainty of 1 mm. There must be some discussion for this definition.

The development of precision measurements with different transition frequencies seems to be useful, even if the uncertainty below 10^{-19} is not attainable with any transition frequencies. For example, the relativistic effect is significant with the motion of electrons in highly charged ions. Precision measurement of the transition frequencies of highly charged ions is useful to confirm the relativistic effects. We do not know in detail about the nuclear structure, which might be clarified from the precision measurement of nuclear transition frequencies. The vibrational-rotational transitions are given by the motion of the nuclei of constituent atoms. Precision measurement of molecular transition frequencies gives detailed information about nuclear mass.

4.4 Laser cooling of positronium

Positronium (Ps) is a bound state of an electron and a positron (Figure 4.17). Its energy structure is much simpler than atoms because it is given only by the electromagnetic force. The lifetime of Ps in the $(n, L) = (1, 0)$ state (n: principal quantum number, L: rotational quantum number) is limited by the pair annihilation in an area where the wavefunctions of the electron and positron overlap. The total spin of an electron and a positron is 1 or 0. In the spin state of 1, there is a magnetic repulsion force between the electron and positron and the lifetime (1.42×10^{-7} s) is much longer than that in the spin state of 0 having the magnetic attractive force (1.25×10^{-10}s).

Here, we introduce a project to obtain cold Ps in the spin state of 1. With the project in the U. of Tokyo [26], Ps was produced by the collision of a positron on the surface of the converter (SiO_2 aerogel), which is cooled to 4 K by a cryogenic refrigerator. The thermalization of Ps to 100 K was attained. However, further thermalization within the lifetime of Ps is difficult because of the small mass of Ps.

Laser cooling using the 1S-2P cycle transition (243 nm) is required for further cooling. The kinetic energy of several tens of K was expected to be attained using a pulsed cooling laser (length of 300 ns) and a frequency bandwidth of 460 GHz (effective for a wide range of velocities). Lower temperature is attained by frequency chirping of the cooling laser: large $\Delta\nu = \nu_0 - \nu_L$ at the beginning to decelerate fast Ps and small $\Delta\nu$ to decelerate slow Ps afterwards. Recently, one-dimensional laser cooling was performed by chirping the frequency within 100 ns, as shown below. The cooling laser is a pulsed laser with a length of 0.1 ns and a frequency bandwidth of 9 GHz. The cooling laser is inserted into a cavity

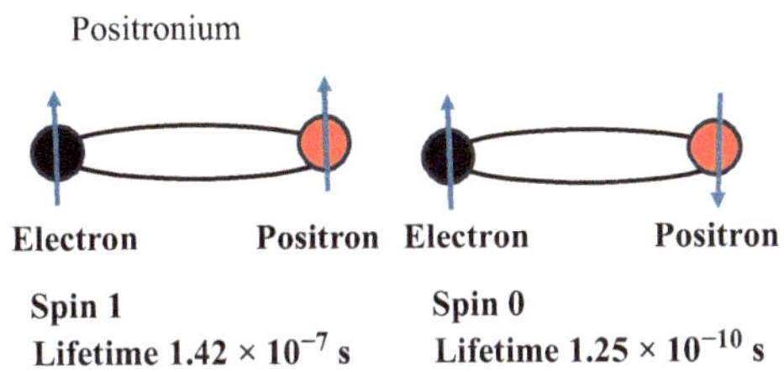

Figure 4.17 Positronium with spin 1 and 0

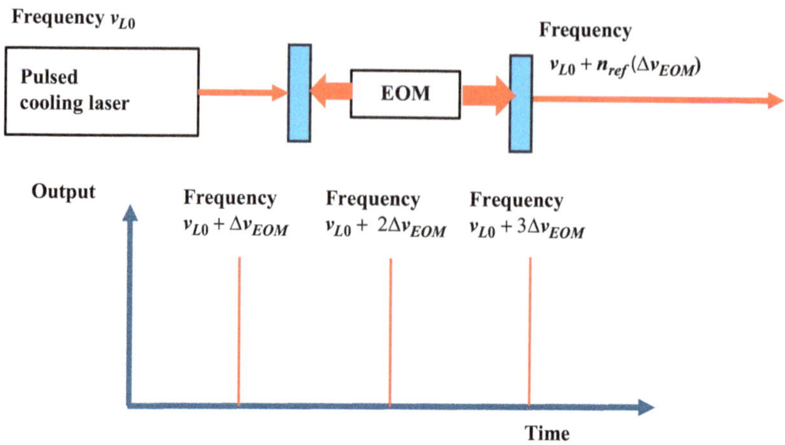

Figure 4.18 Frequency chirping of pulsed cooling laser using a cavity containing an electric-optical modulator (EOM)

containing an electro-optical modulator to give a frequency shift of $\Delta\nu_{EOM}$ with one interaction. Repeating the reflection n_{ref} times in the cavity, the frequency of the output light is shifted by $n_{ref}(\Delta\nu_{EOM})$. Output with a higher reflection time is delayed by $n_{ref}\tau_{ref}$ (τ_{ref}: reflection time in the cavity 4.2 ns) and the frequency of light irradiated to Ps is chirped with a rate of 490 GHz/μs as shown in Figure 4.18. With this method, the kinetic energy below 1 K was obtained [27].

After three-dimensional cooling, the Bose-Einstein condensate (Chapter 7.4) may be realized with Ps.

4.5 Observation of ultra-fast phenomena

The eigenenergies of atoms or molecules are given by the motion of electrons or nuclei, but the energy eigenfunctions at each state have no dependence on time because of the uncertainty principle of time and energy. First, we consider the vibrational motion of diatomic molecules as shown in Figure 4.19. The eigen energy of vibrational motion with the frequency of ν_v is given by $(n_v + 1/2)h\nu_v$ (Chapter 3.8.4). The energy eigenvalues are measured using CW laser with a narrow frequency bandwidth to obtain high frequency resolution, but the vibrational motion cannot be observed with this method. To observe the vibrational motion itself, the periodic change of the density distribution should be observed with a time resolution higher than $1/\nu_v$. Development of ultra-short pulse lasers is useful to increase time resolution. To observe the vibrational motion of N_2 or O_2 molecules, a pulse laser with a length of 10^{-15} is required.

How can we obtain ultra-short-pulse laser light? When the laser oscillates with the cavity (L_c), frequency components of $\nu_{com}(n_m) = n_m\nu_{rep} + \nu_{ceo}$ are resonant to the cavity, where $\nu_{rep}(= c/2L_c)$ is the repetition rate of the pulsed laser and ν_{ceo} is

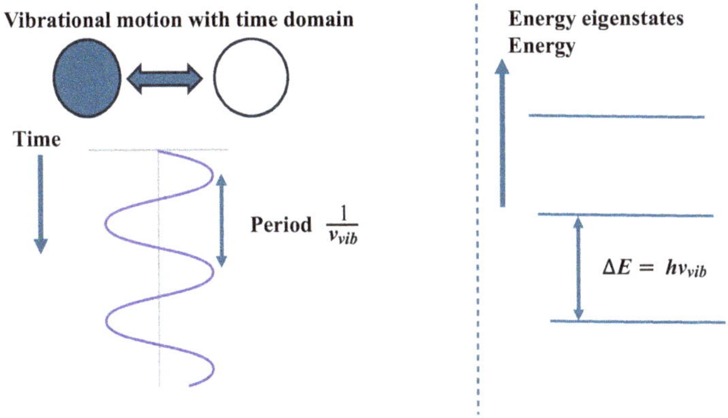

Figure 4.19 Comparison of vibrational motion and the energy eigenvalues

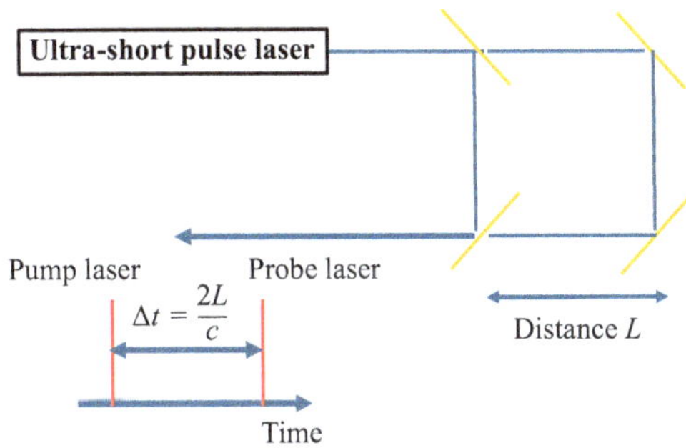

Figure 4.20 Pump-and-probe laser lights with ultra-short pulses. The delay of the probe laser after the pump laser is given by the difference in optical path length.

offset frequency induced by frequency-dependent refractive index (Chapter 4.2.5). When many frequency components ($N_0 - \Delta N_m \leq n_m \leq N_0 + \Delta N_m$) oscillate simultaneously, keeping the phase relation constant (mode locking), the pulse length is given by $1/(\Delta N_m)\nu_{rep}$ [28]. The irradiation of ultra-short pulse lasers is interpreted as the simultaneous irradiation of laser lights with many frequency components. The vibrational motion is observed as the coupled state between different energy eigenstates. The pulse length was on the order of 10^{-12} s at the beginning of the 1980s. In the latter half of the 1990s, pulse lasers with a length of 10^{-14} s have been developed.

The relaxation process (for example, electron spin in a magnetic material) in solids after excitation by a pump laser is monitored using a probe laser, which is irradiated with a delay time of Δt. Pump-and-probe lasers are obtained by dividing an ultra-short pulse of laser light. The delay time Δt is given by the difference in optical path length as shown in Figure 4.20.

Pulse lasers with lengths below 10^{-15} s have been developed recently, which makes it possible to detect the motion of electrons in chemical reactions [29,30]. The mechanism of photosynthesis might be clarified from the observation of electron motion.

The peak intensity of ultra-short pulse laser is high (10^9 W) with low energy. Ultra-short pulse lasers are often used for burning only small areas of an object (useful for medical purpose) because energy from the laser light is concentrated on the target point without expansion of thermal energy to the surroundings.

References

[1] Bauch A. 'The PTB primary clocks CS1 and CS2'. *Metrologia*, 2005; 42 43–54. DOI10.1088/0026-1394/42/3/S06

[2] Vanier J. and Audoin C. *The Quantum Physics Atomic Frequency Standards*. Adam, Hilger, Bristol and Philadelphia. 1989; p 610

[3] Vanier J. and Tomescu C. *The Quantum Physics Atomic Frequency Standards: Recent Developments*. CRC press. 2015; p 1

[4] 'Time and Frequency from A to Z: H'. *Physics Laboratory*. https://www.nist.gov/pml/time-and-frequency-division/popular-links/time-frequency-z/time-and-frequency-z-h NIST. Hydrogen Maser. Retrieved 2010-04-06

[5] Hasegawa A., Fukuda K., Kajita M. *et al.* 'Accuracy evaluation of optically pumped primary frequency standard CRL-O1'. *Metrologia*, 2004; 41 257–263. DOI:10.1088/0026-1394/41/4/006

[6] Shimizu F. 'Stark spectroscopy of NH_3 ν_2 band by 10-μ CO2 and N2O lasers'. *J. Chem. Phys.*, 1970; 52 3572–3578. DOI:10.1063/1.1674111

[7] Johns J. W. C. and McKellar A. R. W. 'Stark spectroscopy with the CO laser: the ν_2 fundamentals of H_2CO and D_2CO'. *J. Mol. Spectrosco.*, 1973; 48 354–371. DOI:10.1016/0022-2852(73)90200-2

[8] Freund S. M., Duxbury G., Römheld M., Tiedje J. T. and Oka T. 'Laser Stark spectroscopy in the 10 μm region: the $\nu3$ bands of CH_3F'. *J. Mol. Spectrosco.*, 1974; 52 38–57. DOI:10.1016/0022-2852(74)90004-6

[9] Job V. A., Patel N. D., Cunha R. D. and Kartha V. B. 'High resolution diode laser measurements on the ν_2 bands of $^{14}NH_3$ and $^{15}NH_3$'. *J. Mol. Spectrosco.*, 1983; 101 48–60. DOI:10.1016/0022-2852(83)90005-X

[10] Haensch T. W. and Shawlow A. L. 'Cooling of gases by laser radiation'. *Opt. Commun.*, 1975; 13 68–69. DOI:10.1016/0030-4018(75)90159-5

[11] Metcalf H. J. and Straten P. van der. *Laser Cooling and Trapping*. New York, Springer-Verlag Inc; 1999

[12] Wineland D. J., Drullinger R. E. and Walls F. L. 'Radiation-pressure cooling of bound resonant absorbers'. *Phys. Rev. Lett.*, 1978; 40 1639–1642. DOI:10.1103/PhysRevLett.40.1639

[13] Dalibard J. and Cohen-Tannoudji C. 'Laser cooling below the Doppler limit by polarization gradients: simple theoretical models'. *J. Opt. Soc. Am. B*, 1989; 6 2023–2045. https://doi.org/10.1364/JOSAB.6.002023

[14] Morigi G., Eschner J., Cirac J. T. and Zoller P. 'Laser cooling of two trapped ions: sideband cooling beyond the Lamb-Dicke limit'. *Phys. Rev. A*, 1999; 59 3797–3808. DOI:10.1103/PhysRevA.59.3797

[15] Hamann S. E., Haycock D. L., Klose G., Pax P. H., Deutsch I. H. and Jessen P. S. 'Resolved-sideband Raman cooling to the ground state of an optical lattice'. *Phys. Rev. Lett.*, 1998; 80 4149–4152. DOI:10.1103/PhysRevLett. 80.4149

[16] Perrin H., Kuhn A., Bouchoule I., Pfau T. and Salomon C. 'Raman cooling of spin-polarized cesium atoms in a crossed dipole trap'. *Euro. Phys. Lett.*, 1999; 46 141–144. DOI:10.1209/epl/i1999-00236-0

[17] Huntemann N., Sanner C., Lipphardt B., Tamm Chr. and Peik E. 'Single-ion atomic clock with 3×10^{-18} systematic uncertainty'. *Phys. Rev. Lett.*, 2016; 116 063001. DOI:10.1103/PhysRevLett.116.063001

[18] Hausser H. N., Keller J., Nordmann T. *et al.* '$^{115}In^{+} - {}^{172}Yb^{+}$ coulomb crystal clock with 2.5×10^{-18} systematic uncertainty'. *Phys. Rev. Lett.*, 2025; 134 023201. DOI:10.1103/PhysRevLett.134.023201

[19] Marshall M. C., Rodriguez Castillo D. A., Arthur-Dworschack W. J. *et al.* 'High-stability single-ion clock with 5.5×10^{-19} systematic uncertainty'. *Phys. Rev. Lett.*, 2025; 135 033201. DOI:10.1103/hb3c-dk28

[20] Recommended values of standard frequencies by Bereau International des Poids et Measures https://www.bipm.org/en/publications/mises-en-pratique/standard-frequencies

[21] Beloy K., Bodine M. I., Bothwell T. *et al.* 'Frequency ratio measurements at 18-digit accuracy using an optical clock network'. *Nature*, 2021; 591 564–569. DOI:10.1038/s41586-021-03253-4

[22] Aeppli A., Kim K., Warfield W., Safronova M. S. and Ye J. 'Clock with 8×10^{-19} systematic uncertainty'. *Phys. Rev. Lett.*, 2024; 133 023401. DOI:10.1103/PhysRevLett.133.023401

[23] King S. A., Spieß L. J., Micke P. *et al.* 'An optical atomic clock based on a highly charged ion'. *Nature*, 2022; 611 43–47. DOI:10.1038/s41586-022-05245-4

[24] Elwell R., Schneider C., Jeet J. *et al.* 'Laser excitation of the ^{229}Th nuclear isomeric transition in a solid-state host'. *Phys. Rev. Lett.*, 2024; 133 013201. DOI:10.1103/PhysRevLett.133.013201

[25] Zhang C., Ooi T., Higgins J. S. *et al.* 'Frequency ratio of the ^{229m}Th nuclear isomeric transition and the ^{87}Sr atomic clock'. *Nature*, 2024; 633 63–70. DOI:10.1038/s41586-024-07839-6

[26] Shu K., Murayoshi T., Fan X. *et al.* 'Study on Bose-Einstein condensation of positronium'. *J. Phys. Conf. Series*, 2017; 791 012007. DOI:10.1088/1742-6596/791/1/012007

[27] Shu K., Tajima Y., Uozumi R. *et al.* 'Cooling positronium to ultralow velocities with a chirped laser pulse train'. *Nature*, 2024; 633 793–797. DOI:10.1038/s41586-024-07912-0

[28] Mayer B., Regler A., Sterzl S. *et al.* 'Long-term mutual phase locking of picosecond pulse pairs generated by a semiconductor nanowire laser'. *Nat. Commun. B*, 2017; 8 15521. DOI:10.1038/ncomms15521

[29] Paul P. M., Toma E. S., Breger P. *et al.* 'Observation of a train of attosecond pulses from high harmonic generation'. *Nature*, 2001; 292 1689–1692. DOI:10.1126/science.1059413

[30] Klausz F. and Ivanov M. 'Attosecond physics'. *Rev. Mod. Phys.*, 2009; 81 163–234. DOI:10.1103/RevModPhys.81.163

Chapter 5
Detecting subtle relativistic effects

Abstract

The theory of special relativity indicates slower time progress in a moving coordinate. The theory of general relativity indicates slower time progress in a gravitational potential area. Although these effects are too slight to be detected with a quartz clock, they were detected using atomic clocks. The change of gravitational potential propagates as a gravitational wave, which induces a change in length with a ratio below 10^{-21}. This effect was detected by laser interferometry. The strict confirmation of $E = mc^2$ is also introduced. The acceleration of charged particles has been performed using a synchrotron, taking the relativistic effect into account. A relativistic effect is also detected from the Sagnac effect.

5.1 Observation of time dilation with atomic clocks

Chapter 3.6 indicates that time progress in a moving frame is slower than in a static frame. This effect was confirmed by observation of particles with short lifetimes in cosmic rays from distant places (several light-years). This is because particles move at high velocities (close to the speed of light), and time progresses more slowly in the coordinates co-moving with the particles. The lifetime of particles in an accelerator becomes longer as they are accelerated [1]. However, this relativistic effect with the velocities below 300 m/s was too small to be detected with the uncertainty of quartz clocks. Chapter 3.7 indicates that time goes slower in a place with a gravitational potential field (gravitational redshift), but this effect, with the difference of the altitude of 10 km, is 10^{-12} and cannot be detected without an atomic clock. Atomic clocks are required to confirm that the changes in the progress of time are given by (3.82) and (3.100).

The change of the progress of time was confirmed by the Hafele-Keating experiment [2]. They took four Cs commercial atomic clocks aboard commercial airliners. They flew twice around the world (once westward and once eastward) to consider the effect of the rotation of the Earth. After returning, they found the disagreement between the clocks on board (average of four clocks) and on Earth (westward 275 ns, eastward −40 ns), which were consistent with the theories of special (slower progress of time by motion) and general (faster progress of time by

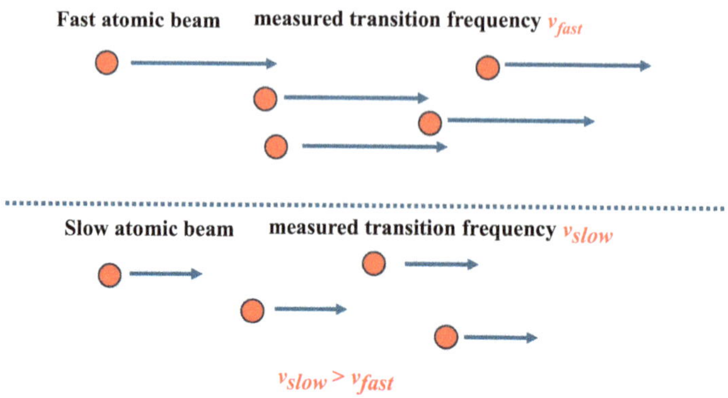

Figure 5.1 Quadratic Doppler shift by relativistic effect. Transition frequency measured with fast atoms is lower than that measured with slow atoms.

high altitude) relativities. This result indicates a combination of two effects. The separate detections are shown as follows.

The transition frequency of atoms moving with velocity v becomes lower than the real transition frequency with a factor of $\sqrt{1-(v/c)^2}$ because of the slower progress of time in a moving frame (called the quadratic Doppler shift). With Cs atomic clocks using atomic beams (Chapter 4.1), the quadratic Doppler shifts of order 10^{-13} can be observed. Using a magnetic selector to localize atoms in one state, only atoms with a limited velocity range contribute to the measurement [3]. This atomic velocity range depends on the structure of the magnet selector. Comparing the atomic clocks with different ranges of atomic velocity, there is a difference between the direct measurements of atomic transition frequencies, as shown in Figure 5.1. Making a correction of the quadratic Doppler shift from a measured velocity distribution, the frequency difference reduces significantly. The quadratic Doppler shift is observed clearly from a comparison between Cs atomic clocks with a thermal atomic beam and those using laser-cooled atoms (quadratic Doppler shift $< 10^{-17}$). The quadratic Doppler shift induced by different velocities in the range of 0–40 m/s was measured precisely using the $^{27}Al^{+}$ transition frequency (attained measurement uncertainty of 10^{-18}), which was in good agreement with the theory of relativity [4]. The quadratic Doppler shift has been analyzed with the approximation $\sqrt{1-(v/c)^2} \approx 1-(v/c)^2/2$. For more detailed analysis, $\sqrt{1-(v/c)^2} \approx 1-(v/c)^2/2-(v/c)^4/8$ should be used. But the term of $-(v/c)^4/8$ has never been observed with the current measurement uncertainty. To detect this term with the velocity of 10^4 m/s, the measurement uncertainty of 10^{-19} is required, which has never been attained.

The gravitational redshift is observed as a change in the transition frequency with changing altitude, as shown in Figure 5.2 (ratio of $g/c^2 = 1.1 \times 10^{-16}$/m, where

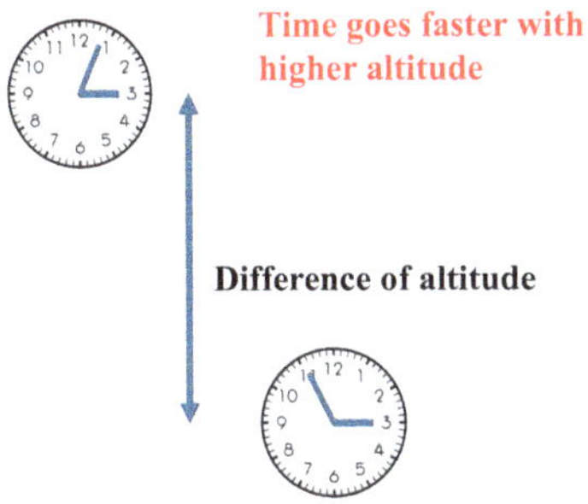

Figure 5.2 Gravitational redshift, with which time goes faster at a higher altitude

g is gravitational acceleration). The gravitational redshift with the difference of altitude of 1000 m can also be detected by a commercial Cs atomic clock. It was measured by comparing the frequencies of Cs atomic clocks located in a valley and on a nearby mountain [5].

The change in the gravitational redshift with the difference in the altitude of 16 cm was observed using the $^{27}Al^{+}$ transition frequency [4]. On the other hand, we can estimate the difference in altitude with the uncertainty of a few cm from the comparison of atomic transition frequencies, which can be measured with the uncertainty of 10^{-18} [6]. The measurement of the change of altitude using atomic clocks might offer new opportunities to explore seismology and volcanology [7]. However, there must be a discussion to give the standard of gravitational potential, because the gravitational potential is not perfectly constant, as shown in Chapter 4.3.4.

Previously, no discrepancies were discovered between the theory of relativity and the observations about the time progression. But we cannot deny the possibility of observing the discrepancy between the theory of general relativity and the detailed observation with ultra-high gravitational potential (around neutron stars?) when it is performed in the future.

5.2 Detection of gravitational waves

As shown in Chapter 3.7, gravity was described as a distortion of space. The change of the gravitational potential propagates as a wave with the speed of light. The gravitational wave induces a change in the space size in the direction perpendicular to the propagation direction (z-direction): repeat of (expand in x-direction + contract in y-direction) and (contract in x-direction + expand in y-direction). It has been difficult to observe the change in size because it is so small, being a ratio below 10^{-21}.

However, a phenomenon was discovered to confirm the existence of gravitational waves indirectly. The orbital period of a binary neutron star PSR 1913 + 16 (7.75 h) decreased by 7.65×10^{-5} s/year [8] (Figure 5.3). They are expected to collapse 300 million years from now. This result shows the loss of energy in the orbital motion. By analogy to charged particles in a circular motion losing kinetic

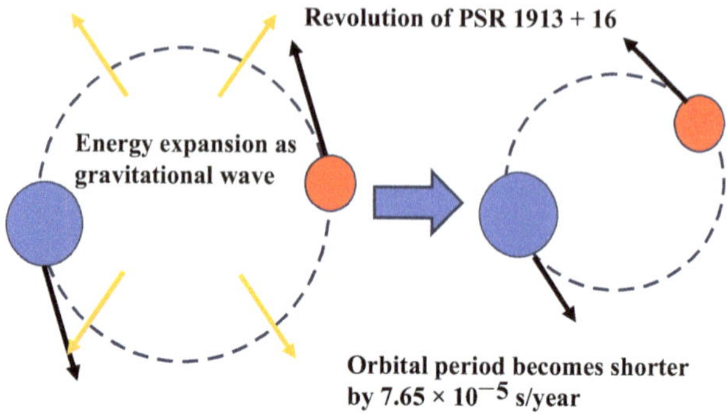

Figure 5.3 Orbital period and radius of the binary neutron star decrease because energy is lost with the expansion of the gravitational wave

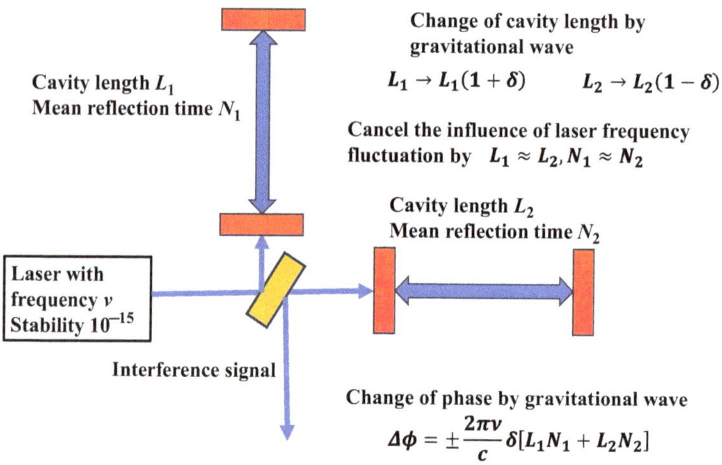

Figure 5.4 Laser interferometry system to detect the gravitational waves

energy radiating the electromagnetic wave (Chapter 3.3.4), it is reasonable to state that objects with large mass in an orbital motion lose kinetic energy radiating the gravitational waves. The measured variation in the orbital period is in good agreement with the estimation from the theory of general relativity to within 0.2%. The accuracy of the atomic clock contributed to measuring the orbital period with high precision (present accuracy of 1.5×10^{-5} s).

A laser interferometer was constructed for the direct detection of the variation of the space size by the gravitational wave. The apparatus is shown in Figure 5.4. A laser light (Nd: YAG laser, frequency $\nu = 2.828$ THz) is split by a half mirror and propagates into two perpendicular directions. In the arms of the interferometer (resonators of length $L_{1,2}$), the light is reflected many times ($N_{1,2}$ times). Both laser beams are brought together in the half mirror again to form an interference pattern exhibiting a phase difference of

$$\phi = \frac{2\pi\nu}{c}(L_1N_1 - L_2N_2). \tag{5.1}$$

When the gravitational wave changes the cavity length by $L_{1,2} \to L_{1,2}(1 \pm \delta)$, there is a change in the phase difference of

$$\Delta\phi = \pm\frac{2\pi\nu}{c}\delta(L_1N_1 + L_2N_2). \tag{5.2}$$

Assuming $L_1 \approx L_2$ and $N_1 \approx N_2$, the influences of the frequency fluctuation of the laser light on the phase progress in both directions are almost the same. Its effect on the interference signal is suppressed by six orders through significant cancellation.

The first direct observation of the gravitational wave succeeded on September 14, 2015, at the Laser Interferometer Gravitational-Wave Observatory (LIGO) in Hanford (Washington, US) and Livingston (Louisiana, US) [9]. The change of cavity length of $\delta L_{1,2}/L_{1,2} < 10^{-22}$ was detected with $L_1 \approx L_2 \approx 4$ km and $N_1 \approx N_2 \approx 500$. The detected waveform (frequency increased from 30 Hz to 250 Hz) matched the prediction of general relativity for a gravitational wave emerging from the merger of a pair of black holes of 36 and 29 solar masses. From the slight difference in the detection time between the signals at two observatories, the source was found to be 1.3 billion light-years distant. This signal was recognized as that of the gravitational wave because the same figures of signals were observed at two positions. The noise level of the signal is dominated by the shot noise of laser light; the flow rate of photons with the average of n_p/s fluctuates within $n_p\left(1 \pm 1/\sqrt{n_p}\right)$/s as shown in Chapter 2.2.1. In LIGO, the ratio of shot noise to the average flow rate has been reduced by increasing the laser power by 2 W → 20 W → 100 W. With high power, the fluctuation of radiation pressure can change the cavity length (Chapter 3.3.3); therefore, power stabilization with 10^{-6} is also required.

On August 14, 2017, the gravitational wave from the merger of black holes of 12 and 7 solar masses, 1 billion light-years distant, was detected at the LIGO and VIRGO interferometers (Cascina, Italy) simultaneously. Its origin has been determined in more detail from the difference in the detection time at distant places [10]. On August 17, 2017, a gravitational wave from two merging neutron stars was detected by LIGO and VIRGO, which was also observed in the electromagnetic radiation [11]. Until 2023, more than 90 gravitational waves have been detected in LIGO. The gravitational wave will also be detected with some other observatories (for example, KAGRA, Japan) in the near future. Observing a gravitational wave at different places, we can get more detailed information about the origin. The gravitational wave detectable on the Earth is limited to the frequency area higher than 10 Hz, because the vibrational noise on the surface makes it difficult to detect it in the lower frequency area.

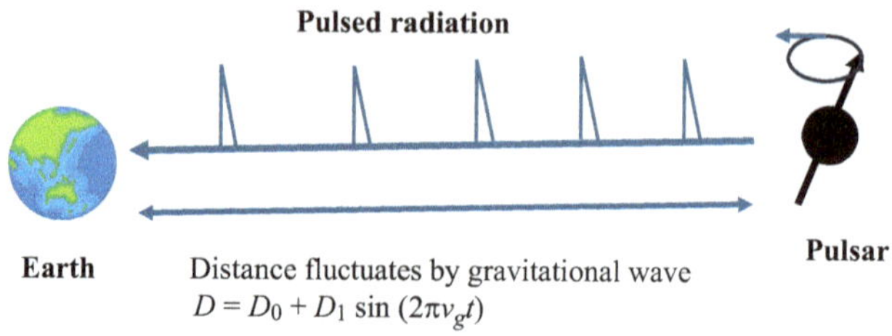

Figure 5.5 The fluctuation of the period of the pulsed radiation from a pulsar was observed because the distance between the pulsar and Earth fluctuates due to gravitational waves

There is a project to detect gravitational waves by the interferometer using cavities in space constructed by mirrors in three satellites, aiming for the detection of gravitational waves with a frequency below 10 Hz in a circumstance free from vibration on the surface of the Earth. The cavity length is much larger than the size of the Earth. It might be possible to detect the gravitational wave soon after the birth of the universe, whose frequency is estimated to be in the order of 1 mHz.

The gravitational wave with ultra-low frequency (1–10 nHz) was detected by three collaborations (the North American group NANOGrav, the European Pulsar Timing Array, and the Parks Pulsar Timing Array in Australia) from the fluctuation of pulsar timing [12]. Pulsars are neutron stars emitting light, radio waves, or X-rays. The emission is observed as a pulse on the Earth with periods between 10^{-3} and 10 s, because the emission direction rotates with the precession of the neutron star. The period to detect the emission from pulsars is generally stable, but recently, some oscillational fluctuation with the pulsar timings (order 100 ns) was discovered. They insist that this result indicates the fluctuation of the distance D between the Earth and the pulsars induced by the background gravitational wave $\left(D = D_0 + D_1 \sin\left(2\pi\nu_g t\right)\right)$ as shown in Figure 5.5. To obtain this result, the measurement of the pulsar timing should be continued for 20 years (period of the gravitational wave is several years). Although the fluctuation of the pulsar timing is only 2–4 times larger than the measurement uncertainty, this conclusion was given by the consistent results from three collaborations.

5.3 Experimental confirmation of $E = mc^2$

The theory of special relativity introduced rest energy $E = mc^2$ (Chapter 3.6). The existence of rest energy has been confirmed by comparing the change in total mass and the release of energy in the nuclear reaction. However, there was a question of whether this relation holds strictly. If this relation were found to be slightly incorrect, the fundamentals of modern physics would be violated.

This relation was confirmed in more detail by comparing the change of mass and the release energy when Si and S nuclei absorb neutrons (Figure 5.6) [13]. With

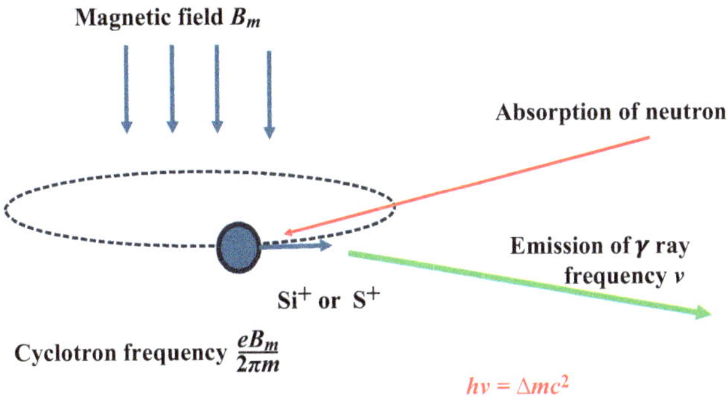

Figure 5.6 Comparison between the change of mass of trapped atomic ion and release energy (frequency of γ-ray)

this nuclear reaction, energy is released as γ-rays. Change of mass was measured from the change of cyclotron frequency (frequency of the circular motion of ion in magnetic field), and release energy was measured by the frequency of emitted γ-rays. Finally, the result

$$1 - \frac{\Delta mc^2}{E} = (-1.4 \pm 4.4) \times 10^{-7} \tag{5.3}$$

was obtained. The relation $E = mc^2$ was confirmed with the highest accuracy.

5.4 Cyclotrons and synchrotrons

A cyclotron is an accelerator of charged particles having a cyclic motion under a magnetic field. Under a magnetic field of B_m, a particle with an electric charge of q_e and mass of m has a cyclical motion with the frequency of

$$\nu_{cyc} = \frac{q_e B_m}{2\pi m} \tag{5.4}$$

with any velocity of the particle. Acceleration is performed by giving an AC electric field with a frequency of ν_{cyc} for acceleration. While the velocity is low enough to apply the Newtonian mechanics, the acceleration can be performed using an AC electric field with a constant frequency. The motion radius with the speed of v_S is given by

$$R_{cyc} = \frac{v_S m}{q_e B_m}, \tag{5.5}$$

which is increased as the particle is accelerated.

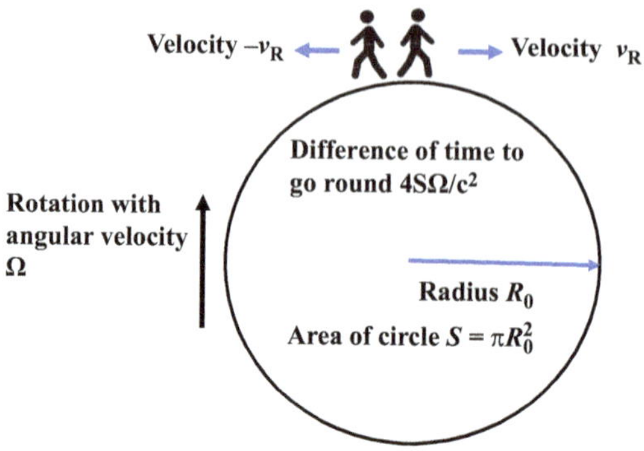

Figure 5.7 The Sagnac effect, which gives the difference in the time to go around the rotating circle because of the relativistic effect

Taking the relativistic effect into account, (5.4) and (5.5) are revised to

$$v_{cyc} = \frac{q_e B_m}{2\pi m}\sqrt{1-\left(\frac{v_S}{c}\right)^2}, \tag{5.6}$$

$$R_{cyc} = \frac{v_S m}{q_e B_m \sqrt{1-(v_S/c)^2}}, \tag{5.7}$$

The acceleration by the cyclotron is no longer effective because the cyclotron motion is not periodic. A synchrotron is an accelerator that accelerates charged particles, keeping the motion radius constant by controlling B_m and the frequency of the AC electric field. The control of these values is consistent with the estimation by the theory of special relativity. Now, many charged particles accelerate to velocities close to the speed of light, and the world of high energy was developed (Chapter 9).

5.5 The Sagnac effect

The Sagnac effect is an effect in which the time to go around a rotating circle (radius R_0, area S) depends on the path because of the relativistic effect, as shown in Figure 5.7. For the motion with the velocity v_R parallel and antiparallel to the rotation, the time to go around the circle is given by [14]

parallel velocity $v_R + R_0\Omega$

time to go around the circle $T_+ = \dfrac{2\pi R_0}{v_R\sqrt{1-\left(\frac{v_R+R_0\Omega}{c}\right)^2}}$, antiparallel velocity $v_R - R_0\Omega$

$$\text{time to go around the circle } T_- = \frac{2\pi R_0}{v_R\sqrt{1-\left(\frac{v_R - R_0\Omega}{c}\right)^2}}, \tag{5.8}$$

and the difference of time is given by

$$\delta T = T_+ - T_- = -\frac{4\pi R_0^2\Omega}{c^2} = -\frac{4S\Omega}{c^2}. \tag{5.9}$$

This time difference is monitored with the interference of a wave with frequency ν propagating in both directions (interference is observed at the propagating half-round). The phase shift is given by

$$\Delta\phi_{Gy} = \frac{1}{2} \times 2\pi\nu(\delta T) = -\frac{4\pi\nu S\Omega}{c^2}. \tag{5.10}$$

This effect is currently observed using ring lasers propagating in both directions. Measurement of angular velocity with this method is important to monitor the position of an airplane, ship, or submarine. Equation (5.8) is not valid for light propagation because the speed of light is always constant. However, the frequency shift by the Doppler effect gives

$$\nu_\pm = \nu\left(1 \pm \frac{R\Omega}{c}\right)$$

$$\Delta\phi_{Gy} = \frac{1}{2} \times \frac{2\pi R}{c} \times 2\pi(\nu_- - \nu_+) = -\frac{4\pi\nu S\Omega}{c^2}, \tag{5.11}$$

which gives the same phase difference as (5.10). Equations (5.10) and (5.11) indicate that the Sagnac effect is monitored with higher sensitivity using a wave with a higher frequency. Now there is a project to measure the angular velocity with higher accuracy, monitoring the Sagnac effect using the interferometry of atomic waves, because the frequency of atomic waves ($\nu = mc^2/h$, where m is atomic mass) is much higher than laser lights (Chapter 7.2).

5.6 Application of the global positioning system (GPS)

Currently, latitude, longitude, and altitude are measured by receiving radio signals from satellites (1.1–1.6 GHz) called the Global Positioning System (GPS). The position measurement is based on the distance between the satellites and the receiver. With the propagation time between the receiver and satellites, 1–3, the distance from each satellite $L_{1-3} = c(T_{1-3} - T_0)$ (T_0 is the time in the receiver, T_{1-3} is the time in satellites 1–3) is determined. Measuring the distances from three satellites 1–3, the position is determined to be two crossing points of spherical surfaces with the distance of L_{1-3} from satellites 1–3 as shown in Figure 5.8. The

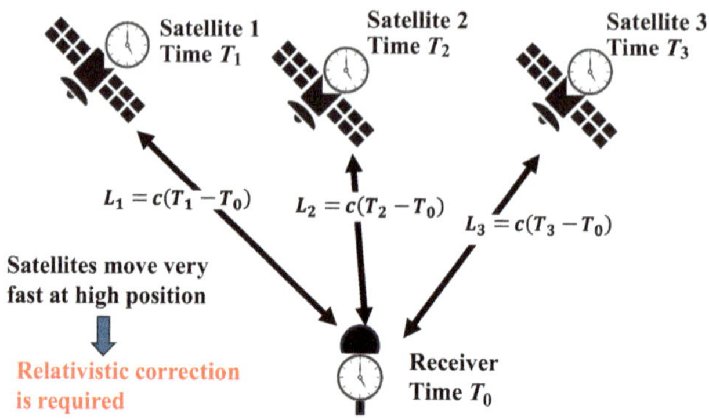

Figure 5.8 *Position measurement by GPS. Distances from four satellites are calculated from the propagation time of the radio wave. Relativistic correction is required for clocks in satellites.*

clocks in the satellites are commercial Cs or ^{87}Rb clocks, with which the uncertainty is on the order of 10^{-13}. But a quartz clock is used for the receiver because it is required to be chip-scale. The accuracy of T_0 is not high enough, and there might be an error of 1 μ s, which corresponds to a positional error of 300 m. Measuring also the distance from satellite 4 $L_4 = c(T_4 - T_0)$, the possible combination of L_{1-4} is limited to one position with the corrected value of T_0.

Currently, there are 24 satellites for fundamental use of GPS geolocation and 7 other satellites for temporal operation in the case of some trouble with fundamental operation. GPS users can receive radio signals from 6 to 10 satellites anywhere on Earth. Receiving signals from many satellites, the accuracy of the position measurement is improved.

The precise measurement of time by an atomic clock makes a great contribution to accurate position measurement. However, it is questionable if the accuracy of position measurement is improved by using atomic clocks with lower uncertainty, because there is a nonzero uncertainty of the speed of radio waves in the air. The speed of radio waves in the air has a slight shift from that in a vacuum (defined value), which has a dependence on the atmospheric pressure and humidity.

Note that the relativistic correction is required for the time comparison between the clocks in the satellites and the receiver. Satellites move with a velocity of 3.8 km/s, which makes the slower progress of time with the ratio of 1.2×10^{-4} s in one day (theory of special relativity). The satellites are at a height of 20,000 km, and the progress of time is faster than that on the surface with the ratio of 1.5×10^{-4} s in one day (theory of general relativity). The total error by both effects is the advance of 3×10^{-5} s in one day, which gives the error of 9 km for position measurement. Relativistic correction is required for precise position measurement [15].

The GPS system was developed by the United States for military use in 1973, aiming for smooth contact between armies or accurate control of missiles, etc. At first, the use of GPS was occupied by the US military. The position of civil aircraft was estimated by measuring the propagation time of radio waves from the basement. There was a position uncertainty of 5 km with this method, and the aircraft faced risks of flying the wrong route. The use of GPS was also opened for civil use after the incident involving Korean Air Flight 007 (1983). However, intentional degradation of the accuracy (SA) was added to civil users' times so that the US military can maintain its advantage over foreign military forces. While SA was operating, the position accuracy was of the order 100 m. In 2000, the US government discontinued the use of SA. The main reason was that other countries were also launching their own satellites, and the United States no longer remained the sole provider of space-based position measurements. Now, an accuracy of a few meters can be given to civilian receivers.

References

[1] Kaminer I., Nemirovsky J., Rechtsman M., Bekenstein R. and Segev M. 'Self-accelerating Dirac particles and prolonging the lifetime of relativistic fermions'. *Nat. Phys.*, 2015; 11 261–267. DOI:10.1038/nphys3196

[2] Hafele J. C. and Keating R. E. 'Around-the-world atomic clocks: predicted relativistic time gains'. *Science*, 1972; 177 (4044) 166–168. DOI:10.1126/science.177.4044.166

[3] Mungall A. G. 'The second order doppler shift in cesium beam atomic frequency standards'. *Metrologia*, 1971; 7 49–56 DOI:10.1088/0026-1394/7/2/002

[4] Chou C. W., Hume D. B., Rosenband T. and Wineland D. J. 'Optical clocks and relativity'. *Science*, 2010; 329 1630–1633. DOI:10.1126/science.1192720

[5] Briatore L. and Leschiutta S. 'Evidence for the earth gravitational shift by direct atomic-time-scale comparison'. *Nuovo Cim B.*, 1977; 37B 219–231. DOI:10.1007/BF02726320

[6] Takano T., Takamoto M., Ushijima I. *et al.* 'Geopotential measurements with synchronously linked optical lattice clocks'. *Nature Photonics*, 2016; 10 662–666. DOI:10.1038/nphoton.2016.159

[7] Tanaka Y. and Katori H. 'Exploring potential applications of optical lattice clocks in a plate subduction zone'. *J. Geodesy*, 2021; 95 93 1–14. DOI:10.1007/s00190-021-01548-y

[8] Taylor J. H. and Weisberg J. M. 'A new test of general relativity: gravitational radiation and the binary pulsar PS R 1913+16'. *Astrophys. J.*, 1982; 253 908–920.

[9] Abbott B. P., Abbott R., Abbott T. D. *et al.* 'Observation of gravitational waves from a binary black hole merger'. *Phys. Rev. Lett.*, 2016; 116 061102. DOI:10.1103/PhysRevLett.116.061102

[10] Abbott B. P., Abbott R., Abbott T. D. *et al.* 'GW170814: a three-detector observation of gravitational waves from a binary black hole coalescence'. *Phys. Rev. Lett.*, 2017; 119 141101. DOI:10.1103/PhysRevLett.119.141101

[11] Drout M. R., Piro A. L., Shappee B. J. *et al*. 'Light curves of the neutron star merger GW170817/SSS17a: implications for r-process nucleosynthesis'. *Science*, 2017; 358 1570–1574. DOI:10.1126/science.aaq0049

[12] Castelvecchi D. 'Monster gravitational waves spotted for first time'. *Nature*, 2023; 619 13–14. DOI:10.1038/d41586-023-02167-7

[13] Rainville S., Thompson J. K., Myers E. G. *et al*. 'A direct test of E=mc^2'. *Nature*, 2005; 438 1096–1097. DOI:10.1038/4381096a

[14] Sagnac G. 'L'ether lumineux demontre par l'effet du vent relatif d'ether dans un interferometre en rotation uniforme'. *Comptes Rendus*, 1913; 157 708–710.

[15] Feraoni V. *Special Relativity*. Spring Science & Business Media; 2013; p 54.

Chapter 6

Expanding the observation frontier in astronomy

Abstract

Until the nineteenth century, astronomical research had been performed only with optical observation in a small fraction of the Milky Way. Observation outside of the Milky Way started in the twentieth century. Observation was also performed with radio waves, infrared, ultraviolet, and X-rays. The universe is not immutable, and it has been in the process of expansion since the Big Bang 13.8 billion years ago. The expansion is accelerated by dark energy, whose identity is a mystery of modern physics. The gravitational force between galaxies is much larger than the estimation considering just visible matter, which indicates the existence of dark matter. To search for the identity of dark matter, many experiments (for example, searching for variation in fundamental constants) have been performed.

6.1 Is the universe immutable?

There was a paradox assuming the universe with an infinite size and an immutable structure. We consider the total light intensity from all stars in the universe. Taking the intensity of light emitted from a star Γ_{star}, the intensity of this light on the Earth is proportional to Γ_{star}/R^2 (R: distance between the star and the Earth). If the stars are distributed uniformly with the density of ρ_{star} in $R_{\min} < R < R_{\max}$, the total intensity is

$$I_{tot} \propto 4\pi \int_{R_{\min}}^{R_{\max}} \rho_{star} \frac{\Gamma_{star}}{R^2} R^2 dR = 4\pi \rho_{star} \Gamma_{star} (R_{\max} - R_{\min}). \tag{6.1}$$

Taking $R_{\max} \to \infty$, the total light intensity is infinite. It cannot be dark at night. Therefore, the universe cannot be immutable with infinite size.

Until the nineteenth century, the universe was believed to be immutable, because the sight of stars does not change in our lives. Note that we could see only small fractions of stars in the Milky Way.

The theory of general relativity indicates that gravity is the distortion of the coordinates of space (Chapter 3.7). The universe cannot be given by a straight coordinate system while it contains massive objects. The size of the universe must be reduced, and all matters should be merged in the future. Einstein could not accept the conclusion from his own theory and introduced a cosmological term to derive the immutable universe.

This paradox was solved when galaxies outside the Milky Way were observed, and Hubble's law was discovered in 1929 [1]. This law states that the objects observed in deep space are moving away from the Earth at a velocity approximately proportional to their distance from the Earth. The motion velocities of stars are estimated by analyzing the radiation emanating from stars, which is intense or weak at some discrete frequencies. The observed frequency distribution indicates the spectrum of materials in the star. The observed spectrum is shifted by the Doppler effect, from which we can estimate the motion velocity of the star. The distances to the far distant stars are estimated by the brightness of supernova explosions because the annual parallax is observed only for stars closer than 1000 light years. Assuming the uniform release energy of a supernova explosion, its brightness is inversely proportional to the square of the distance.

Lights from distant stars are observed as infrared or microwave radiation because of the Doppler effect, with the velocity proportional to the distance. The total light intensity from the whole universe is finite, and it is dark at night. Considering that the present universe is in the process of expansion, there is no discrepancy with the theory of general relativity. Einstein canceled the introduction of the cosmological term. Observing the galaxies outside the Milky Way, the astronomical recognition changed perfectly.

6.2 Multi-wavelength observations

Until the beginning of the twentieth century, the observation was performed only with the optical region. Observing astronomical radiation in different frequency areas, we can observe many phenomena that we cannot see in the optical region. The research of radio wave astronomy started in 1933, aiming at the observation of the densest part of the Milky Way in the constellation of Sagittarius [2]. Radio astronomy has led to substantial increases in astronomical knowledge, particularly with the discovery of several classes of new objects, including pulsars, quasars, and radio galaxies. Such objects represent some of the most energetic physical processes in the universe. The spatial resolution of radio waves is low because of long wavelengths. To improve the resolution, the interferometry of signals from several observatories at distant places is required. The role of the atomic clock is important to guarantee simultaneous observations at each observatory.

The observation of radiation in the infrared, ultraviolet, and X-ray areas is also useful, but observation on the Earth is difficult for these frequency areas because of scattering or absorption in the air. Observations using balloons or rockets have been performed, but they are now performed mainly using satellites.

6.3 The Big Bang model for the birth of the universe

Hubble's law indicates that the objects observed in deep space are moving away from the Earth with a velocity approximately proportional to their distance from the Earth. With the time reversal of the motion, the universe converges to one point at 13.8 billion years ago, as shown in Figure 6.1. The universe was born by the explosion at one point, called the "Big Bang." This idea was supported by the discovery of the cosmic microwave background (CMB) in 1965 [3]. The frequency distribution of the CMB corresponds to the blackbody radiation with a thermodynamic temperature of 2.725 K in all directions within the accuracy of 10^{-5}. The uniform temperature shows that the universe was a single object in the past. The discovery of the CMB was the greatest result of radio wave astronomy. The temperature of the universe during the Big Bang was unimaginably high. Soon after the Big Bang (10^{-36}–10^{-34} s), there was a phase transition from a high-energy vacuum to a low-energy vacuum and an expansion was induced with a speed much faster than the speed of light (called inflation) [4,5]. Without inflation, the uniform structure of the universe cannot be explained. The theory of relativity prohibits the motion of an object faster than the speed of light, but it does not prohibit the expansion of space itself faster than the speed of light.

Protons and neutrons were constructed by the binding of quarks (Chapter 9.3) within 10^{-5} s. There were both particles and antiparticles (Chapter 8.1). Antiparticles were annihilated via the collision with particles within 10 s (the size of the universe 10 light years). Particles remained in the universe also after all antiparticles were annihilated because their number slightly exceeded that of antiparticles with the ratio of (10^9 +1: 10^9). Helium nuclei were constructed via the fusion of protons and neutrons within several minutes. Ten minutes after the Big Bang (size of the universe: 1000 light-years), the fusion of nuclei stopped because of the decreased temperature. No other nuclei were constructed until the birth of stars.

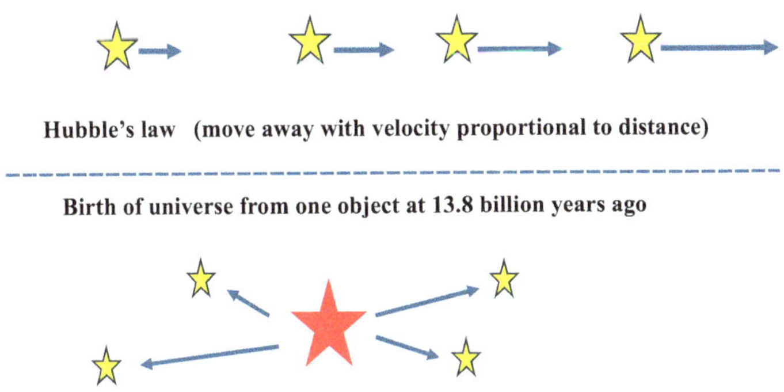

Figure 6.1 Hubble's law indicating that objects move away with a velocity proportional to distance. With time reversal, all matters are merged as one object. The universe was born with an explosion called the "Big Bang."

After 380,000 years, the temperature decreased to around 3000 K, and atoms were formed from nuclei and electrons. After this period, the universe became transparent to electromagnetic waves, and the CMB is understood to be the radiation from this period.

The procedures shown above are all theoretical predictions. We cannot confirm them from the observation, except that the polarization of the CMB is consistent with the inflation model.

Since the Big Bang model was established, the observation of quasars at long distances became a hot topic for astronomers aiming to search for sights in the past. Previously, a quasar at 13.1 billion light-years distant was observed, from which we can see the sight at 13.1 billion years ago [6]. The sight before the transparent radiation of the universe cannot be observed with the electromagnetic wave but might be observed with the gravitational wave. Its frequency is expected to be much lower than that observed previously (Chapter 5.2) because of the Doppler effect by the expansion of the universe. Aiming at the detection of the gravitational wave with a frequency around 1 mHz, there is a project to construct a laser interferometry system using cavities with a length larger than the size of the Earth, constructed by mirrors in satellites. Observing the gravitational wave, detailed information about the inflation model might be obtained.

6.4 Formation of the galaxies and stars

Soon after the discovery of CMB, its temperature distribution was believed to be uniform. However, if the universe were perfectly uniform, the formation of galaxies and stars should not be possible. However, the fluctuation in temperature distribution of the CMB at levels of 10^{-6} was clarified after the improvement of measurement accuracy [7]. The inhomogeneous polarization of the CMB was also discovered. This temperature fluctuation derives from the quantum fluctuation in the density of matter in the early universe, and it was imprinted shortly after the Big Bang.

The procedure for forming galaxies and stars has not been clarified perfectly. The most advantageous hypothesis insists that the matters (dominated by dark matter shown in Chapter 6.6) are accumulated by the gravitational interaction in the area where the initial densities are higher than the surroundings. Supermassive black holes (Chapter 3.7) were discovered at the center of many galaxies. We do not know whether the supermassive black holes were formed in the galaxies or the galaxies were formed around the supermassive black holes. A supermassive black hole was also discovered in a galaxy that is 13.1 billion light-years distant, indicating that supermassive black holes were formed within 700 million years after the Big Bang.

6.5 Dark energy

It is reasonable to expect that the velocity of expansion decreases due to gravitational interaction. However, precision measurements of the positions and velocities of galaxies show that the expansion of the universe has been accelerating for the past 6.62 billion years. The force causing this acceleration is called "dark energy" [8]. The identity of dark energy is a mystery of modern physics. The cosmological term, which Einstein once introduced and later canceled, is introduced again to describe the present expansion of the universe. Detailed measurement of the gravitational wave might give useful information to find the identity of the dark energy. The detailed information of the nonhomogeneity of the CMB also gives information about the inflation soon after the Big Bang, which gives a chance to solve the mystery of dark energy (we do not know if the inflation was caused by the same force as dark energy).

Many researchers have been eager to search for the identity of dark energy. On the other hand, a new hypothesis was given to deny the existence of dark energy. Lee *et al.* insist that the estimation of the distance to the stars from the brightness of the supernova explosion is questionable [9]. The release of energy by a supernova explosion might have a dependence on the circumstance. With their corrected estimations of distances, no acceleration was indicated. However, more discussion will be required about the reliability of the corrected distance.

6.6 Dark matter

The gravitational interaction in the universe is much larger than that estimated from the mass of visible matter, as shown below. In 1933, Zwicky assumed the existence of dark matter to explain the mass of galaxies and to describe the velocity of galaxies in a galaxy cluster [10]. He observed the relative velocity between galaxies in the Coma cluster to be 1000 km/s. The size of the Coma cluster was estimated to be 3.3 million light-years. The mass of clusters was estimated to be 3×10^{14} M_{sol} (M_{sol}: solar mass), so that the gravitational force is balanced with the centrifugal force. This estimation is two orders larger than the total mass of visible objects, as shown in Figure 6.2. The same result was

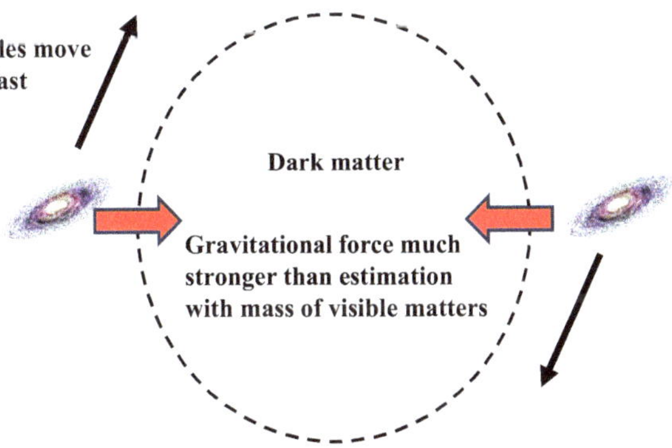

Figure 6.2 There must be dark matter so that the gravitational force balances the centrifugal force with a fast relative velocity

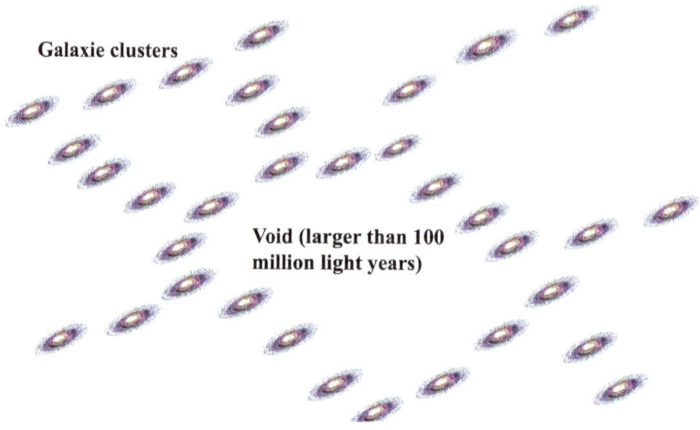

Figure 6.3 Distribution of galaxy clusters appears to have a bubble-like structure

obtained by observation of other galaxy clusters. However, Zwicky mentioned not only the existence of dark matter but also the possibility of violation of the Newtonian gravitational law at large distances. The existence of dark matter is determined from the velocity of H atoms at the edges of galaxies [11]. The measured velocity is much higher than expected, and the gravitational force required to balance the centrifugal force is ten times larger than that given by the visible matter within each galaxy.

In the 1980s, the distribution of galaxy clusters was mapped, and a bubble-like structure of the universe appeared [12] along with areas called voids, where there are no galaxies (Figure 6.3). The sizes of these voids are larger than 100 million light-years. To explain this structure, the total mass of dark matter should be much larger than that of visible matter. There was a hypothesis that galaxies cannot be formed without dark matter. However, a galaxy without dark matter, NGC1052-DF_2, was discovered [13], and a new model may be required to clarify the mechanics of forming galaxies without dark matter.

There have been some arguments that the mystery of gravitational force can be explained by a modification of Newtonian gravitational law, but the modified gravitational law could not explain the structure of galaxies discovered afterward. The final proof for the existence of dark matter is that the mass distribution measured from the bend of the optical path (Chapter 3.7) is quite different from that of visible matter [14].

Several candidates have been considered for the identity of dark matter. The rotation of polarization of CMB by the interaction with dark matter has been observed with a reliability of 99.2% [15]. This phenomenon indicates the violation of the CP symmetry (Chapter 8.3). There could be more than one answer about the identity of dark matter, but it is expected to include some particle that induces the violation of CP symmetry. Some mysteries of violating CP symmetry (non-equal abundance of particles and antiparticles) might be solved from the identity of dark matter.

Weak interactive massive particles (WIMPs) are one of the candidates. This is the lightest supersymmetric particle with a neutral electric charge. The mass was theoretically estimated to be 1–100 m_p. (m_p: proton mass). This particle has been considered to have a weak interaction (Chapters 8.3 and 9.3) because violation of CP symmetry is observed [15]. Experimental efforts to detect WIMPs have been performed by (i) searching for expected products at the annihilation of WIMPs (γ-ray, neutrino, etc.) in nearby galaxies and galaxy clusters; (ii) searching for the collision of WIMPs with nuclei in the laboratory; and (iii) direct production of WIMPs in colliders. However, this particle was never discovered until 2024, and it cast doubt on the simplest WIMP hypothesis. Just recently (2025), γ-ray from the halo of the Milky Way was discovered with the peak frequency of 5×10^{24} Hz (20 GeV). Its residual mapping indicates the spherically symmetric distribution of halo components. Therefore, the observed γ-ray is consistent with the one emitted at the pair annihilation of a WIMP with the mass of $500 - 800 m_p$ [16]. In the near future, we will see whether it is concluded as the first discovery of WIMP.

Axion with zero spin and mass of $10^{-11} - 10^{-9}$ m_e (m_e: electron mass) is also a candidate for dark matter, although it has never been discovered yet. The broadening of the wavefunction is the same order as the size of a galaxy because of the small mass; therefore, axions are expected to be in the Bose-Einstein condensate state (uniform phase of wavefunction, see Chapter 7.4) also at room temperature. There is an effort to detect axions from the variation of fundamental constants (Chapter 6.7.3). Axion is a particle predicted to solve the strong CP problem (Chapter 9.6).

Some heavenly bodies can also be candidates of dark matter. For example, black holes cannot be observed with light and can be one candidate for dark matter. However, their total mass is estimated to be lower than 40% of the total mass of dark matter. Neutron stars with weak light or planets, which have never been observed, might also be candidates.

6.7 Searches for variations in fundamental constants

Physical laws have been established with many fundamental constants. With different values of the fundamental constants, the appearance of the universe would be quite different. With a larger value of the fine structure constant $\alpha = e^2/2\varepsilon_0 hc$, atoms with heavy nuclei could not exist because of the repulsive force between protons. With smaller values of α, molecular bonding is not possible because of a weaker electric force. We would not exist with any other value of α. The combination of suitable fundamental constants looks like too much of a coincidence. If fundamental constants have a dependence on time and position, we may understand that we are living in an epoch with suitable combinations of fundamental constants ($\alpha = 0.007297$); the idea that they arose by chance is not required. In 1937, Dirac mentioned for the first time the possibility of time-varying fundamental constants (not the requirement) [17]. However, the change in fundamental constants was expected to be too small to detect with the measurement uncertainties in the twentieth century. Currently, some transition frequencies can be measured with uncertainties lower than 10^{-16} and the search for such variations has become an active topic of investigation. We are curious, particularly if the current physical laws were also valid soon after the birth of the universe.

When there is a variation in a fundamental constant X, there is a variation in the energy structure of atoms and molecules. The transition frequencies change with different ratios. The variation in X (ΔX) is estimated from the variation in the ratio of two transition frequencies $\nu_1(\propto X^{K_{X1}})$ and $\nu_2(\propto X^{K_{X2}})$,

$$\frac{\Delta(\nu_1/\nu_2)}{(\nu_1/\nu_2)} = (K_{X1} - K_{X2})\frac{\Delta X}{X}. \tag{6.2}$$

A nonzero ($\Delta X/X$) is detected when $\Delta(\nu_1/\nu_2) > \delta(\nu_1/\nu_2)$, where $\delta(\nu_1/\nu_2)$ is the measurement uncertainty in (ν_1/ν_2). It is advantageous to measure the ratio between two transition frequencies with large $K_{X1} - K_{X2}$ and low $\delta(\nu_1/\nu_2)$. It is not simple to satisfy both requirements, because the combination of transition frequencies with high $|K_{X1} - K_{X2}|$ is often sensitive to the fluctuation of circumstances.

The variation in fundamental constants for a long period is monitored from a comparison of (ν_1/ν_2) values for a quasar and for the Earth. The appearance of a quasar at 10 billion light-years signifies a time 10 billion years ago. With observations from quasars in different directions, we can distinguish between spatial and temporal variations. With this method, we cannot measure the quick variation of fundamental constants. The other method is to search for variation in (ν_1/ν_2) for several years by the precision frequency measurement in a laboratory (Chapter 4). The oscillational variations in fundamental constants are detected only with measurements in a laboratory. We, however, cannot distinguish between a spatial variation and a temporal one from the measurement in a laboratory because the solar system moves in the universe with a velocity of 240 km/s (one order faster than Earth's orbital velocity).

6.7.1 *The fine structure constant*

Estimating the sensitivity parameter of the fine structure constant α is not simple. While the relativistic effects with electron motion in atoms are negligibly small, the transition frequencies between different electron energy states are proportional to $m_e\alpha^2$. Comparing the transition frequencies proportional to $m_e\alpha^2$, the ratio is also constant when there is a variation in α. For atoms with heavy nuclei, the motion velocity of electrons is fast, and relativistic effects induce the higher-order term of α as shown in (3.180). The variation in α is detected from the variation in the ratio between transition frequencies of heavy and light atoms.

The astronomical search for the variation in α has been performed, comparing the spectra of several kinds of metal ions in quasars and the laboratory sample spectra. Several different results have been reported, and many efforts have been made to reduce measurement uncertainties. Recent observations indicate almost null results, as shown below. The variation in the fine structure constant at

13 billion years ago was obtained to be $\Delta\alpha/\alpha = (-2.18 \pm 7.27) \times 10^{-5}$ by observing the spectra of Si^+, Al^+, Fe^+, and Mg^+ ions in the quasar J1120+0641 [18]. Observing the Fe^+ spectrum in the quasar J110325-264515 (10 billion light-years distance), $\Delta\alpha/\alpha = (1.56 \pm 1.78) \times 10^{-6}$ was obtained [19]. There might be variation in physical laws soon after the Big Bang, but it might be difficult to discover the wreckage from the appearance after the era of transparent radiation.

Since the atomic transition frequencies were measured with the uncertainty below 10^{-16}, a search for the variation in α has been performed by comparing the following transition frequencies measured in laboratories.

(1)

$$^{199}\mathrm{Hg} + {}^2\mathrm{S}_{1/2} \rightarrow {}^2\mathrm{D}_{5/2}\ (K_\alpha = -3.2) \text{ and } {}^{27}\mathrm{Al}^+\ {}^1\mathrm{S}_0 \rightarrow {}^3\mathrm{P}_0\ (K_\alpha = 0.008)$$

$$(d\alpha/dt)/\alpha = (-1.6 \pm 2.3) \times 10^{-17}\ (/\text{year})[20],$$

(2)

$$^{171}\mathrm{Yb}^+\ {}^2\mathrm{S}_{1/2} \rightarrow {}^2\mathrm{F}_{7/2}(K_\alpha = -6.0) \text{ and } {}^{171}\mathrm{Yb}^+\ {}^2\mathrm{S}_{1/2} \rightarrow {}^2\mathrm{D}_{3/2}\ (K_\alpha = 1.0)$$

$$(d\alpha/dt)\alpha = (-2.0 \pm 2.0) \times 10^{-17}\ (/\text{year})\ [21]$$

$$(d\alpha/dt)\alpha = (-0.5 \pm 1.6) \times 10^{-17}\ (/\text{year})[22],$$

$$(d\alpha/dt)\alpha = (1.0 \pm 1.1) \times 10^{-18}\ (/\text{year})[23],$$

The variation rate in α is below 10^{-18}/year. Precision measurement of transition frequencies of highly charged ions or nuclear transition frequencies is advantageous to observe the variation in α because of $|K_\alpha| > 30$. When these transitions are measured with the uncertainty of 10^{-18} (seems possible in the near future, as shown in Chapter 4.3.3), we will see whether a nonzero variation in α is discovered or a null result is confirmed with lower uncertainty.

6.7.2 *The proton–electron mass ratio*

If there is a variation in the fine structure constant α, there should also be a variation in the proton-to-electron mass ratio $\mu_{pe}(= m_p/m_e)$. The proton mass is two orders larger than the total mass of the constituent particles (two up-quarks and one down-quark). The kinetic energy of quarks dominates the rest energy of protons, as shown in Chapter 9.3. The proton mass changes drastically with the variation in the electromagnetic force between quarks ($\propto \alpha$). Reference [24] indicates the relation,

$$\frac{\Delta\mu_{pe}}{\mu_{pe}} = R_{\text{var}}\frac{\Delta\alpha}{\alpha}, \tag{6.3}$$

where R_{var} is a constant between 28 and 40, given by the details of the Grand Unification Theory (GUT). Comparing the variations in α and μ_{pe}, we can obtain useful information for GUT.

It is difficult to search for the variation in μ_{pe} from the comparison between the atomic transition frequencies in the optical region because of $K_{\mu_{pe}} < 10^{-4}$ (reduced mass between electron and nucleus is almost equal to m_e). Precision measurements of molecular vibrational rotational transition frequencies are useful to detect the variation in μ_{pe}, because they are given by the motion of the nucleus. Values of $K_{\mu_{pe}}$ are -0.5 and -1 for the vibrational and rotational transition frequencies, respectively.

An astronomical search for the variation in μ_{pe} was performed from the H_2 molecular transition frequencies. The quasar source J2123-005 (12 billion light-years distance) observed by two different telescopes gave the results of $\Delta\mu_{pe}/\mu_{pe} = (8.3 \pm 4.2) \times 10^{-6}$ and $(5.6 \pm 6.2) \times 10^{-6}$, respectively [25]. A value of $\Delta\mu_{pe}/\mu_{pe} = (0 \pm 1) \times 10^{-7}$ was obtained from the measurement of the CH_3OH transition frequency in a quasar, PKS1830-211 (8 billion light-years distance) [26].

Attainable measurement uncertainty of 10^{-18} has been theoretically estimated for several molecular vibrational transition frequencies [27–29]. Precision measurement of molecular transition frequencies is useful to detect the variation in μ_{pe}, but it has not been realized at the time of 2025. Currently, $(d\mu_{pe}/dt)\mu_{pe}$ has been estimated from a comparison between the Cs hyperfine transition frequency $\left(\propto \mu_{pe}^{-1}\right)$ and the $^{171}Yb^{+}$ S-F transition frequency. Results of $(-0.5 \pm 1.6) \times 10^{-16}$/year [21], $(0.2 \pm 1.1) \times 10^{-16}$/year [22] and $(-0.8 \pm 3.6) \times 10^{-17}$/year [23] were obtained. It is not realistic to reduce the measurement uncertainty of the Cs hyperfine transition frequency to below 10^{-17}. After the realization of precision measurement of molecular transition frequencies with the uncertainty of 10^{-18}, a nonzero variation in μ_{pe} might be detected in the future.

6.7.3 *Axion searches via oscillational variations in constants*

Although null results of the linear drift of fundamental constants have been given, there is still a possibility of detecting an oscillational variation, which might give important information about axions (one candidate for dark matter). The axion is expected to be a particle with zero spin and mass m_ϕ below $10^{-6}\ m_e$, then the axion is in the Bose-Einstein condensate (uniform phase of wavefunction, as shown in Chapter 7.4). The rest energy is given by $m_\phi c^2$ and the wavefunction with a uniform phase is given by

$$\Phi \propto \cos\left[2\pi\nu_{DM}t\right]$$

$$\nu_{DM} = \frac{m_\phi c^2}{h}. \tag{6.4}$$

The interaction with an axion can induce oscillational variation in the fundamental constants, whose frequency indicates the mass of the axion. The change in the density of the axion is observed by the change in the amplitude. Figure 6.4 assumes that the change of density within the period of oscillation is negligible. A more complicated analysis is required when this assumption is not valid.

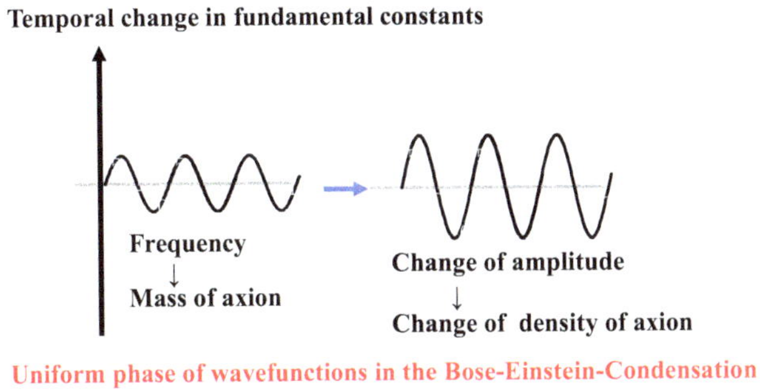

Figure 6.4 *The oscillational variation of fundamental constants induced by interaction with an axion. The frequency indicates the mass of the dark matter particle. The change in the amplitude indicates the change in the density of dark matter.*

A certain measurement time τ_{mes} is required to reduce the measurement statistic uncertainty. Oscillational variation is detected only when the oscillation period is much longer than the measurement time. Therefore,

$$\nu_{DM}\tau_{mes} \ll 1$$

$$m_\phi \ll \frac{h}{c^2\tau_{mes}} \tag{6.5}$$

is required to observe the oscillational variation of fundamental constants. The experiments to observe the oscillational variations in fundamental constants are in progress in several groups [29–33], but the oscillational variation has never been observed previously. The upper limit of their amplitude was determined with a wide range of ν_{DM}. In a lower frequency area, the determined upper limit is lower because the statistical uncertainty is reduced by taking a longer measurement time.

6.8 Discovery of exoplanets

There are planets in the solar system; therefore, it is natural to believe that there are also planets revolving around other stars (exoplanets). However, they were believed to be difficult to observe the reflection of light from the stars.

The first exoplanet around a sun-like star (51 Pegasi b) was discovered in 1995, and its mass is near that of Jupiter [34]. The revolution is not exactly around the star but around the center of mass, as shown in Figure 6.5. Therefore, there is also a revolution motion of the star around the center of mass, which is observed by the periodic change of the Doppler shift of the radiation from the star. Now, the transit method is commonly used; observe the periodic change of the brightness of the star as shown in Figure 6.5. Until July 2024, more than 5500 exoplanets were discovered.

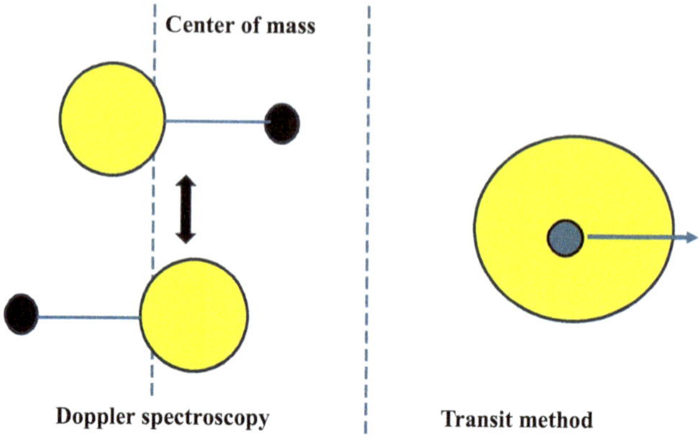

Figure 6.5 *Two methods to discover exoplanets. One is the Doppler spectroscopy method: observing the motion of a star around the center of mass from the periodic change of Doppler shift of radiation from the star. The other is the transit method: observing the periodic change of brightness of the star.*

By measuring exoplanet sizes (diameters) and masses, we can see compositions ranging from rocky (like Earth and Venus) to gas-rich (like Jupiter and Saturn). Some rocky planets may be dominated by water or ice, while others are dominated by iron or carbon.

We are particularly curious about the possibility of finding an exoplanet where some living organisms inhabit. We call it an inhabitable zone mainly from the temperature, where water can be liquid. Previously, several hundred exoplanets were in the habitable zone.

The discovery of exoplanets became possible by observing slight changes in the Doppler shift or brightness of the radiation from stars through precision measurements.

References

[1] Hubble E. 'A relation between distance and radial velocity among extra-galactic nebulae'. *Proceedings of the National Academy of Sciences*, 15 (3) 168–173. DOI:10.1073/pnas.15.3.168

[2] Jansky K. J. 'Radio waves from outside the solar system'. *Nature*, 1933; 132 66. DOI 10.1038/132066a0

[3] Penzias A. A. and Wilson R. W. 'A measurement of excess antenna temperature at 4080 Mc/s'. *Astrophys. J.*, 1965; 142 419–421. DOI:10.1086/148307

[4] Sato K. 'First-order phase transition of a vacuum and the expansion of the Universe'. *Mon. Notice R. Astron. Soc.*, 1981; 195 467–479. DOI:10.1093/mnras/195.3.467

[5] Guth A. H. 'Inflationary universe: a possible solution to the horizon and flatness problems'. *Phys. Rev. D*, 1981; 23 347–356. DOI:10.1103/PhysRevD.23.347

[6] Wang F.,Yang J., Fan X. *et al.* 'A luminous quasar at redshift 7.642'. *Astrophys. J. Lett.*, 2021; 907 L1. DOI:10.3847/2041-8213/abd401

[7] Smooth G. F., Bennett, C. L., Kogut A. *et al.* 'Structure in the COBE differential microwave radiometer first-year maps'. *Astrophys. J. Lett.*, 1992; 396 L1–L5. DOI:10.1086/186504

[8] Perlmutter S., Turner M. S., and While M. 'Constraining dark energy with type ia supernovae and large-scale structure'. *Phys. Rev. Lett.*, 1999; 83 670–673. DOI:10.1103/PhysRevLett.83.670

[9] Lee Y. W., Chung C., Kang Y. and James J. M. 'Further evidence for significant luminosity evolution in supernova cosmology'. *Astrophys. J.*, 2020; 903 22. DOI:10.3847/1538-4357/abb3c6

[10] Zwicky F. 'Die Rotverschiebung von extragalaktischen Nebeln'. *Helvetica Physica Acta*, 1933; 6 110–127

[11] Rubin V., Thonnard N. and Ford W. K. Jr. 'Rotational properties of 21 SC galaxies with a large range of luminosities and radii, from NGC 4605 (R=4kpc) to UGC 2885 (R=122kpc)'. *Astrophys. J.*, 1980; 238 471–487

[12] Gregory S. A. and Thomson L. A. 'The Coma/A1367 supercluster and its environs'. *Astrophys. J.*, 1978; 222 784–799

[13] van Dokkum P., Danieli S., Cohen Y. *et al.* 'A galaxy lacking dark matter'. *Nature*, 2018; 555 629–632. DOI:10.1038/nature25767

[14] Aubourg E., Bareyre P., Bréhin S. *et al.* 'Evidence for gravitational microlensing by dark objects in the Galactic halo'. *Nature*, 1993; 365 623–625. DOI:10.1038/365623a0

[15] Minami Y. and Komatsu E. 'New extraction of the cosmic birefringence from the planck 2018 polarization data'. *Phys. Rev. Lett.*, 2020; 125 221301. DOI:10.1103/PhysRevLett.125.221301

[16] Totani T.'20 GeV halo-like excess of the Galactic diffuse emission and implications for dark matter annihilation'. *J. Cosmol. Astro. Phys.*, 2025;11 080. DOI:10.1088/1475-7516/2025/11/080

[17] Dirac P. A. 'The cosmological constants'. *Nature*, 1937; 139 323. DOI:10.1038/139323a0

[18] Michael R., Wilczynska M. R., Webb J. K., Bainbridge M. *et al.* 'Four direct measurements of the fine-structure constant 13 billion years ago'. *Sci. Adv.*, 2020; 6 (17) 1–9 DOI:10.1126/sciadv.aay9672

[19] Le T. D. 'Wavelengths of [Fe II] from quasar J110325-264515 for a study of space-time variations in the fine-structure constant'. *Result Phys.*, 2019; 12 1035–1037. DOI:10.1016/j.rinp.2018.12.086

[20] Rosenband T., Hume D. B., Schmidt P. O. *et al.* 'Frequency ratio of Al+ and Hg+ single-ion optical clocks; Metrology at the 17th decimal place'. *Science*, 2008; 319 1808–1812. DOI:10.1126/science.1154622

[21] Huntemann N., Lipphardt B., Tamm Chr., Gerginov V., Weyers S. and Peik E. 'Improved limit on a temporal variation of m_p/m_e from comparisons of Yb^+

and Cs atomic clocks'. *Phys. Rev. Lett.*, 2014; 113 210802. DOI:10.1103/PhysRevLett.113.210802

[22] Godun R. M., Nisbet-Jones P. B. R., Jones J. M. *et al.* 'Frequency ratio of two optical clock transitions in $^{171}Yb^{+}$ and constraints on the time variation of fundamental constants'. *Phys. Rev. Lett.*, 2014; 113 210801. DOI:10.1103/PhysRevLett.113.210801

[23] Lange R., Huntemann N., Rahm J. M. *et al.* 'Improved limits for violations of local position invariance from atomic clock comparisons'. *Phys. Rev. Lett.*, 2021; 126 011102. DOI:10.1103/PhysRevLett.126.011102

[24] Calmet X. and Fritzsch H. 'The cosmological evolution of the nucleon mass and the electroweak coupling constants'. *Euro. Phys. J. C.*, 2002; 24 639–642. DOI:10.1007/s10052-002-0944-4

[25] van Weerdenburg F., Murphy M. T., Malec A.L., Kaper L. and Ubachs W. 'First constraint on cosmological variation of the proton-to-electron mass ratio from two independent telescopes'. *Phys. Rev. Lett.*, 2011; 106 180802. DOI:10.1103/PhysRevLett.106.180802

[26] Bagdonaite J., Jansen P., Henkel C., Bethlem H. L., Menten K. M. and Ubachs W. 'A stringent limit on a drifting proton-to-electron mass ratio from alcohol in the early universe'. *Science*, 2013; 339 46–48. DOI:10.1126/science.1225307

[27] Kajita M., Gopakumar G., Abe M., Hada M. and Keller M. 'Test of m_p/m_e changes using vibrational transitions in N_2^{+}'. *Phys. Rev. A*, 2014; 89 032509. DOI:10.1103/PhysRevA.89.032509

[28] Kajita M. 'Accuracy estimation of the $^{16}O_2^{+}$ transition frequencies targeting the search for the variation in m_p/m_e'. *Phys. Rev. A*, 2017; 95 023418. DOI:10.1103/PhysRevA.95.023418

[29] Hannnecke D. 'Optical clocks based on molecular vibrations as probes of variation of the proton-to-electron mass ratio'. *Quantum Sci. Technol.*, 2021; 6 014005. DOI 10.1088/2058-9565/abc863

[30] Hees A., Guéna J., Abgrall M., Bize S. and Wolf P. 'Searching for an oscillating massive scalar field as a dark matter candidate using atomic hyperfine frequency comparisons'. *Phys. Rev. Lett.*, 2016; 117 061301. DOI:10.1103/PhysRevLett.117.061301

[31] Kobayashi T., Takamizawa A., Akamatsu D. *et al.* 'Search for ultralight dark matter from long-term frequency comparisons of optical and microwave atomic clocks'. *Phys. Rev. Lett.*, 2022; 129 241301. DOI:10.1103/PhysRevLett.129.241301

[32] Oswald R., Nevsky A., Vogt V. *et al.* 'Search for dark-matter-induced oscillations of fundamental constants using molecular spectroscopy'. *Phys. Rev. Lett.*, 2022; 129 031302. DOI:10.1103/PhysRevLett.129.031302

[33] Barontini G., Blackburn L., Boyer V. *et al.* 'Measuring the stability of fundamental constants with a network of clocks'. *EPJ Quantum Technol.*, 2022; 9 (12) 1–52 DOI:10.1140/epjqt/s40507-022-00130-5

[34] Mayor M. and Queloz D. 'A Jupiter-mass companion to a solar-type star'. *Nature*, 1995; 378 355–359. DOI:10.1038/378355a0

Chapter 7
Ultra-low-temperature physics

Abstract

Until the nineteenth century, the thermodynamic temperature below 10 K was not attained. Superfluidity and superconductivity were discovered after liquid helium was produced. The kinetic temperature below 1 mK was attained after the laser cooling of gaseous atoms was developed. With ultra-low temperature, the wave nature of atoms became significant. Atomic interferometry was observed, and it is applicable for the measurement of a slight shift in potential energy. With ultra-low kinetic energy, the atomic collision cross-section is much larger than that at room temperature. Bose-Einstein condensate and Fermi degeneracy were observed with ultra-low kinetic energy.

7.1 Superfluidity and superconductivity

The thermodynamic temperature below 10 K was not attained until the beginning of the twentieth century. In 1908, liquid helium with 4 K was produced by the procedure of adiabatic compression (then the temperature becomes higher), cooling by contact with the surroundings (<50 K), and adiabatic expansion. The temperature of liquid helium can be reduced further by reducing the vapor pressure of the helium gas. Producing liquid helium-4, new phenomena, superfluidity, and superconductivity were discovered.

Superfluidity is the phenomenon where a liquid flows without friction on any surface, which allows it to continue to circulate over obstructions. This phenomenon is observed with helium-4 with the temperature below 2.17 K [1,2]. In the 1950s, Vinen performed experiments to indicate the existence of quantized vortex lines in superfluid helium [3]. In the 1960s, Rayfield and Reif confirmed the existence of quantized vortex rings [4].

A helium-4 atom is a boson particle (no limit to the number of particles in a single quantum state). Superfluidity is a phenomenon of a Bose-Einstein condensate (BEC) (Chapter 7.4) where all atoms are condensed to the lowest energy state with a uniform phase of wavefunction. Superfluidity was believed not to be possible with helium-3, which is a fermion particle (no multiple particles in a single quantum state). But it was also attained with helium-3 at a temperature of 2.5 mK, and it is understood as the result of forming a boson particle by pairing two atoms with inverse spin (Cooper pair).

The electric resistance of metals is proportional to the cube of the thermodynamic temperature, because the resistance is caused by the scattering of electrons by the lattice vibration of atoms. Superconductivity is a phenomenon where the electric resistance vanishes perfectly at a temperature lower than a critical temperature. In 1911, the first superconductivity was discovered with Hg (critical temperature 4.19 K), Nb (critical temperature 9.22 K), and Al (critical temperature 1.20 K) [5]. There is no magnetic field inside the superconductor except for the area around the surface (called the Meissner effect [6]). The penetration depth is called "London penetration depth," which is on the order of 100 nm for most materials. Superconductivity is the BEC state of electrons forming Cooper pairs.

Superconductivity is useful to make a large electric current or a high magnetic field. However, attainable electric current is limited by the saturation electric current. As the electric current in a superconductive wire is possible only on the surface, the saturation current is proportional to the radius, while it is proportional to the cross-section for a conventional wire.

Materials with superconductors with higher critical temperatures have been developed. In 2020, CH_8S under the pressure of 267 GPa was found to be superconductive with the critical temperature of 287.7 K [7].

7.2 Wave properties of atoms

Since laser cooling (Chapter 4.2.4) was developed, gaseous atoms with temperatures below 1 mK have been obtained. All particles have wave natures (de Broglie waves) with the wavelength (called the de Broglie wavelength) of

$$\lambda_{dB} = \frac{h}{|\vec{p}|} = \frac{h}{m|v|} = \frac{h}{\sqrt{mk_B T}}, \tag{7.1}$$

where h is the Planck constant, m is the mass, k_B is the Boltzmann constant, and T is the thermodynamic temperature. Equation (7.1) shows that the de Broglie wavelength becomes larger with smaller mass and lower temperature. The masses of an electron and ^{40}Ar atom are 9.11×10^{-31} and 6.68×10^{-26} kg, respectively. With $T = 300$ K, the de Broglie wavelength of an electron is 1.1×10^{-8} m, which is larger than the atomic size. But it is 1.1×10^{-11} m for ^{40}Ar atom, which is smaller than the radius of the electron orbit. Therefore, the wave nature is not significant for atoms at room temperature. When T is reduced to below 1 mK, the wavelength of the ^{40}Ar atom is larger than 6×10^{-9} m, and the phenomena as waves are significant. The expansion of the de Broglie wavelength shows that the position uncertainty of the particles becomes significant as the uncertainty of the momentum becomes smaller (Chapter 3.8.3).

One of the most fundamental characteristics of quantum mechanics is that all two indistinguishable phenomena must exhibit interference. The distribution of

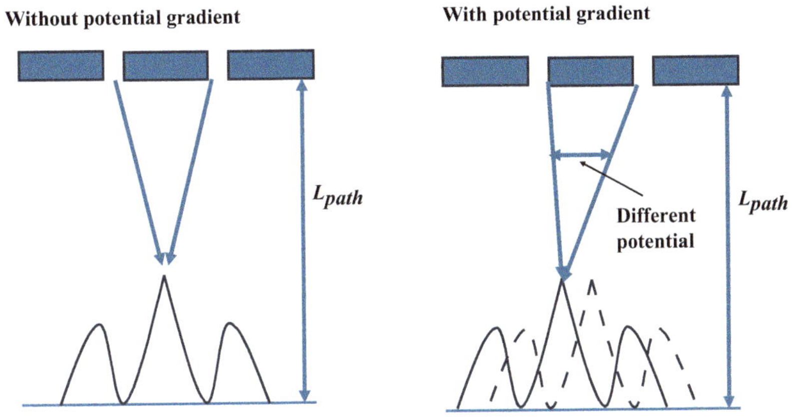

Figure 7.1 Interferometry of atoms passing through a double slit

atoms after passing through a double slit shows an interference as shown Figure 7.1 [8]. Atomic interference is useful not only to confirm the wave nature of atoms but also to detect the slight difference in the potential energies in both paths. With the slight difference of the potential energy ΔP, the phase difference between the two paths is given by

$$\Delta\phi_{\text{int}} = \frac{2\pi\Delta P}{h}\frac{L_{path}}{v_{atom}}, \tag{7.2}$$

where L_{path} is the distance between the double slit and screen and v_{atom} is the atomic velocity. The phase is sensitive to the slight difference in the potential energy with small v_{atom}. The energy shift induced by the electric field was measured with cold Ne atoms with this method [9].

Atomic interference has been applied, for example, to measure the gravitational acceleration with an uncertainty of 3×10^{-9} [10].

Here, we consider the case where the relative position of two particles is given by $\vec{r}$. When two particles are of the same species and in the same quantum state, the relative position $\vec{r}$ and $-\vec{r}$ are not distinguishable. Therefore, there is interference between both wavefunctions, and the following transform is required.

$$\Phi(\vec{r}) \rightarrow \frac{\Phi(\vec{r}) + \Phi(-\vec{r})}{\sqrt{2}} \quad \text{for Boson particles (total spin is integer)}$$

$$\Phi(\vec{r}) \rightarrow \frac{\Phi(\vec{r}) - \Phi(-\vec{r})}{\sqrt{2}} \quad \text{for Fermion particles (total spin is half integer).} \tag{7.3}$$

Considering with $\vec{r} = 0$ for Boson particle, $\sqrt{2}\Phi(0)$ and both particles tend to have the same quantum state. For fermion particles, it is zero. Therefore, two fermion particles cannot take the same quantum state. The discussion with $\vec{r} = 0$ is valid within the overlap of the wavefunctions of two particles $\vec{r} < \lambda_{dB}$. This phenomenon is significant at low temperature (large λ_{dB}).

7.3 Cold collisions

The collision rate is given by

$$\Gamma_{col} = n_a v_{rel}\, \sigma_{col}, \tag{7.4}$$

where n_a is the atomic density, v_{rel} is the mean relative velocity, and σ_{col} is the collision cross-section.

First, we consider a semiclassical treatment. Assuming the classical straight-line track with the closest distance (impact parameter) b_{im}, the collision cross-section of the $a \to b$ transition (for elastic collision, $a = b$) is given by (see Figure 7.2) [11]

$$\sigma_{col} = 2\pi \int b S_{a,b}(b_{im}) db_{im},$$

$$S_{a,b}(b) = \left(\sin \delta_{a,b}\right)^2$$

$$\delta_{a,b} = \frac{2\pi}{h}\int \phi_b^* V(r)\phi_a dt = \frac{2\pi}{h}\int \phi_b^* V(r)\phi_a \frac{dr}{v_{rel}}\quad r = \sqrt{b_{im}^2 + (v_{rel}t)^2}, \tag{7.5}$$

where $\phi_{a,b}$ denote the wavefunctions of atomic internal states a and b.

For the collision with low kinetic energy, the broadening of the de Broglie wavelength is not negligible. The angular momentum of the relative motion is given by $(hL/2\pi) = b_{im}v_{rel}$, which is a discrete value with quantum mechanical treatment because L is an integer. The wavefunctions with the angular momentum

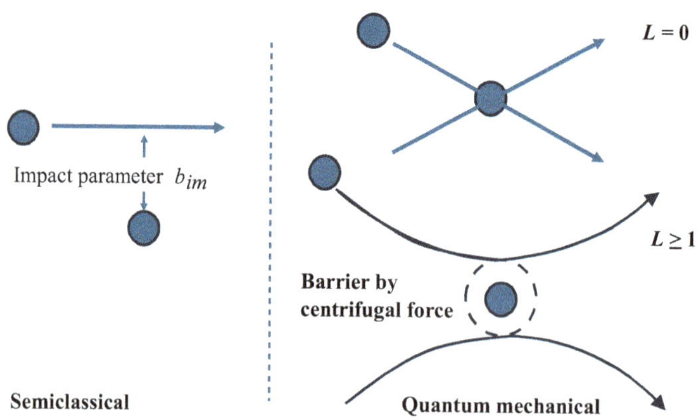

Figure 7.2 Semiclassical and quantum mechanical treatment of atomic and molecular collision

quantum numbers before and after the collision L and L' are given by $\Phi_{L.a}(\vec{r})$ and $\Phi_{L',b}(\vec{r})$. Formulas of $\Phi_{L,a}(\vec{r})$ are obtained by expanding the plane traveling wavefunction $\exp(2\pi i x/\lambda_a)$, where λ_a is the wavelength of the wavefunction and the momentum is given by $p = h/\lambda_a$. With quantum mechanics, the collision cross-section is given by

$$\sigma_{col} = \frac{\lambda_a^2}{\pi} \sum (2L+1) S_{a,b}(L, L'). \tag{7.6}$$

When the interatomic potential works at $r < d$, $\sigma_{col} = \pi d^2$ is roughly obtained from semiclassical treatment (7.5), which is valid for the collision at room temperature. With the ultra-low kinetic energy, λ_a is much larger than d and $\sigma_{col} = \lambda_a^2/\pi$ is roughly obtained, which is much larger than the value obtained from the semiclassical estimation. With ultra-low kinetic energy, the collisional terms with $L \geq 1$ are generally small, because there is a barrier of the centrifugal force potential ($\propto L(L+1)/r^2$) as shown in Figure 7.2. With low kinetic energy, minimum interatomic distance is much larger than the distance where the interatomic potential works.

For the collision between identical atoms in the same state, the treatment shown in (7.3) is required, and (7.6) should be changed to

$$\sigma_{col} = \frac{2\lambda_a^2}{\pi} \sum_{L=even} (2L+1) S_{a,b}(L, L') \text{ Boson}$$

$$\sigma_{col} = \frac{2\lambda_a^2}{\pi} \sum_{L=odd} (2L+1) S_{a,b}(L, L') \text{ Fermion}. \tag{7.7}$$

The collision term with $L = 0$ does not exist for the collision between identical fermion atoms, and the collision effect is suppressed with the ultra-low kinetic energy. According to the description in Chapter 7.2, two identical atoms cannot exist within a distance of λ_{dB} and there is no collision. This effect has been experimentally observed with ^{83}Kr atoms [12]. However, the elastic collision induced by the electric dipole-dipole interaction (potential $\propto \vec{r}^{-3}$) is not suppressed also for fermion particles (polar molecules) [13].

7.4 Bose-Einstein condensates and Fermi degeneracy

The Boltzmann distribution shown in Chapter 3.4 is derived from the probability distribution assuming independent events. This assumption is not valid when atoms are cooled to ultra-low temperatures with high densities. Atomic waves are broader than the interatomic distance, and there is interference between overlapped atomic waves, as shown in Chapter 7.2. Thus, the quantum state cannot be an independent event. Boson particles tend to have the same quantum states.

As shown in Chapter 3.4, the population in an energy state ϵ_E is proportional to $\Omega(\epsilon_E)\exp\left(-\frac{\epsilon_E}{k_B T}\right)$. This relationship also holds using the chemical potential μ_C as follows.

$$\Omega(\epsilon_E)\exp\left(-\frac{\epsilon_E}{k_BT}\right) \to \Omega(\epsilon_E)\exp\left(-\frac{\epsilon_E+\mu_C}{k_BT}\right). \tag{7.8}$$

Here, we consider the distribution of the number of particles in each state n_m. For bosons, n_m can take all integers. For a given temperature, the average of n_m at each energy state ϵ_E is given by

$$(n_m(\epsilon_E))_{ave} = \frac{\Omega(\epsilon_E)\sum_{n=0}^{\infty} n_m \exp\left[-\frac{n_m(\epsilon_E+\mu_C)}{k_BT}\right]}{\sum_{n=0}^{\infty}\exp\left[-\frac{n_m(\epsilon_E+\mu_C)}{k_BT}\right]} = \frac{\Omega(\epsilon_E)}{\exp\left[\frac{(\epsilon_E+\mu_C)}{k_BT}\right]-1}. \tag{7.9}$$

The chemical potential μ_C is a parameter to derive the constant total matter number N_m as follows:

$$N_m = \int \frac{\Omega(\epsilon_E)}{\exp\left[\frac{(\epsilon_E+\mu_C)}{k_BT}\right]-1} d\epsilon_E. \tag{7.10}$$

Here, $\mu_C \geq 0$ is required so that $(n_m(0))_{ave} > 0$ is satisfied in addition to $\epsilon_E \to 0$. For an ultra-low temperature, (7.10) cannot be satisfied with any value of μ_C because of

$$N_{m0} = \int \frac{\Omega(\epsilon_E)}{\exp\left[\frac{(\epsilon_E+\mu_C)}{k_BT}\right]-1} d\epsilon_E < \int \frac{\Omega(\epsilon_E)}{\exp\left[\frac{\epsilon_E}{k_BT}\right]-1} d\epsilon_E < N_m. \tag{7.11}$$

Bosons with a number $(N_m - N_{m0})$ are condensed to the lowest energy. This phenomenon is called a Bose-Einstein condensate (BEC). The phase of the matter waves in the BEC is uniform, and the group treats a single mechanical entity using a wave function on a macroscopic state. Here, we discuss the conditions for obtaining the BEC state in free space. For a given momentum p_q ($q = x, y, z$), the number of states of the position in the region $0 < q < L$ is $\frac{Lp_q}{h} = \frac{L}{\lambda_{dB}}$, where λ_{dB} is the de Broglie wavelength. The number of states with the momentum between p_q and $p_q + dp_q$ is given by $\Omega(p_q)dp_q = \frac{L}{h}dp_q$. Considering the state of the number with the three-dimensional kinetic energy,

$$\Omega(p)dp = \Omega(p_x)\Omega(p_y)\Omega(p_z)\mathrm{d}p_x\mathrm{d}p_y\mathrm{d}p_z = \left(\frac{L}{h}\right)^3(4\pi p^2)dp$$

$$\Omega(\epsilon_E)d\epsilon_E = \frac{4\pi}{h^3}V(2m)^{\frac{3}{2}}\sqrt{\epsilon_E}d\epsilon_E$$

m; mass V: volume

$$N_{m0} < \int \frac{\Omega(\epsilon_E)}{\exp\left[\frac{\epsilon_E}{k_BT}\right]-1} d\epsilon_E = \frac{4\pi}{h^3}V(2mk_BT)^{\frac{3}{2}}\int \frac{\sqrt{x}}{\exp[x]-1}dx. \tag{7.12}$$

Using

$$\int \frac{\sqrt{x}}{\exp(x)-1}dx = \frac{\sqrt{\pi}}{2} \times 2.612, \tag{7.13}$$

the BEC state ($N_m > N_{m0}$) is obtained when

$$\frac{N_m}{V}\left(\frac{h^2}{4\pi m k_B T}\right)^{\frac{3}{2}} > 2.612. \tag{7.14}$$

As shown in Chapter 7.1, superfluidity is given by the BEC state. In 1995, the first BEC state with gaseous atoms was made with Rb atoms in a magnetic trap with the kinetic energy 0.17 μK [14]. In 1998, BEC was attained with Na atoms in an optical trap [15]. If an axion with a mass below $10^{-6}m_e$ exists as dark matter (considered to be a candidate), it is in the BEC state at room temperature shown in Chapter 6.6. In the BEC state, superfluidity is attained, and it does not give any friction to the motion of stars.

In the atomic BEC, the dielectric constant can be very large because of the high atomic density. The speed of light propagating in BEC slowed down to 17 m/s [16].

The BEC state is stable (no phase transition to the liquid or solid) when the interatomic potential is repulsive. When it is switched to attractive (by switching the magnetic field, etc.), the size of the BEC becomes smaller rapidly. Then the uncertainty of the atomic momentum increases, from the uncertainty principle between the momentum and the position. Then the collapse of BEC stops and expansion starts because the expansion by momentum uncertainty overcomes the attractive force. The number of atoms decreases in this procedure. This phenomenon is called "Bose -Nova" [17]. After the expansion of a BEC, a small BEC remains, just like a neutron star is made after the burst of a star. Therefore, research using atoms in the BEC state gives useful information about astronomy.

Particles with the spin quantum number of a half-integer are fermions, with which only one particle can exist in a quantum state (Pauli exclusion principle). The simplest example is electrons in atoms; only one electron can take the state given by the principal quantum number n, rotational quantum number L, magnetic quantum number M_L, and electron spin quantum number M_S. In the 1S state, only two electrons can exist. For a Li atom, the third electron must be in the $n = 2$ state. For the group of fermion particles, energy distribution cannot be localized in the lowest state even with ultra-low temperature, but they can occupy the states in order from the lowest. The distribution in each energy state with the Boltzmann distribution, BEC state, and Fermi degeneracy is shown in Figure 7.3.

However, fermion atoms can pair up with opposite spins (total spin 0) and have a boson-like state (Cooper pair). Paired fermion atoms can also undergo the BEC state. The BEC state of paired fermion particles is related to the superfluidity of ^3He (Chapter 7.1). Superconductivity is also a BEC state of paired electrons. The motion of all particles in the BEC state is macroscopic, and any scatterings of small fractions of particles are forbidden, even if there are impurities.

It is useful to observe the phenomena of atoms in the BEC state trapped by a standing wave of laser light, because it gives important information for solid

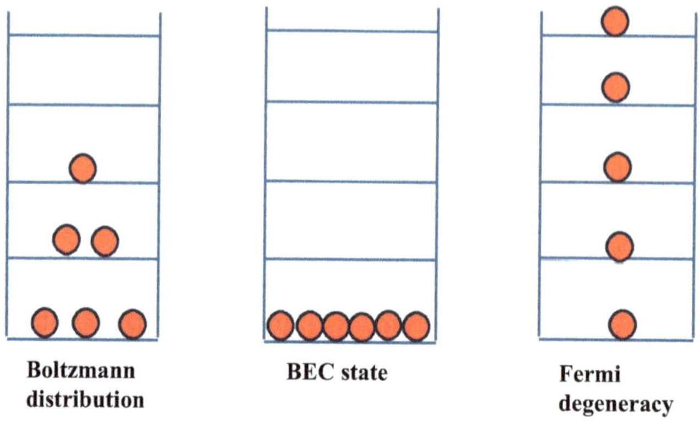

Figure 7.3 *Distribution in each quantum state with Boltzmann distribution, BEC state, and Fermi degeneracy*

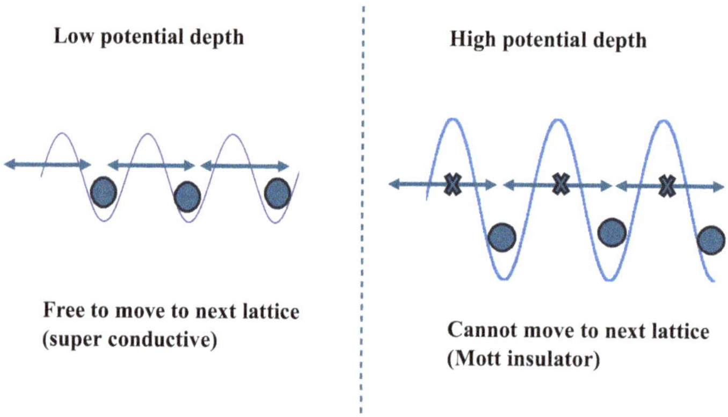

Figure 7.4 *Phase transition between the superfluidity and the Mott insulator*

physics. The electric field of laser light gives a periodic potential field by nodes and antinodes, which is called an "optical lattice." The trap laser frequency is required to be far detuned from the transition frequencies so that it does not induce atomic transitions. The optical lattice gives an analogy of the periodic potential for the electrons in a solid crystal. Note that the optical potential is strictly periodic, while there are lattice defects in a solid crystal. Therefore, we can analyze the properties of electrons in a solid crystal with an ideal model. It is also another merit that we can control the potential depth for free. With low potential depth, atoms are free to move to the neighboring lattice. This is the superfluidity with atoms in the BEC state, which gives an interpretation of the superconductivity of a solid crystal. Increasing the potential depth, there is a phase transition to a state called "Mott insulator," in which atoms cannot move to another lattice because of the repulsive interatomic interaction, as shown in Figure 7.4. With this mechanism, some conductive materials can be transformed into insulators at low temperatures.

The observation of a Cooper pair of fermion atoms (like ^6Li) is useful to search for the superconductivity of a solid crystal. There is research trapping ^{174}Yb (Boson) and ^6Li atoms (fermion) simultaneously in an optical lattice. The ^6Li atoms in an optical lattice have the role of electrons in a crystal, and Yb atoms have the role of the impurity [18].

There is a phenomenon called "Band insulator" with fermion atoms [19]. Atoms cannot move to the neighboring lattice, which is occupied by another atom. This phenomenon looks like Mott insulation, but it is caused not by the interatomic repulsive interaction but by the Pauli exclusion principle prohibiting multi-fermion atoms in the same state.

7.5 Quantum entanglement

One of the most special quantum phenomena is the existence of the "superposition" of multiple states. Now we discuss the superposition of the two eigenstates of two particles A and B

$$\begin{aligned} \Phi_A &= c_1\phi_1 + c_2\phi_2 \\ \Phi_B &= c_\alpha\phi_\alpha + c_\beta\phi_\beta. \end{aligned} \tag{7.15}$$

By the state measurement, the superposition is transformed to one of eigenstates with the possibility of $|c_x|^2$ ($x = 1, 2, \alpha, \beta$). With independent events, the total state is given by

$$\Phi_{tot} = \Phi_A\Phi_B = c_1c_\alpha\phi_1\phi_\alpha + c_1c_\beta\phi_1\phi_\beta + c_2c_\alpha\phi_2\phi_\alpha + c_2c_\beta\phi_2\phi_\beta. \tag{7.16}$$

However, there is also a state called "entangled state"

$$\Phi_{ent} = C_{1\alpha}\phi_1\phi_\alpha + C_{2\beta}\phi_2\phi_\beta. \tag{7.17}$$

Measuring the state of A from the entangled state, we know the state of B simultaneously, even if particle B is at a far distant place from where the propagation time with the speed of light is long. Entangled state between two particles makes it possible to realize the quantum teleportation, transform of a quantum state of a particle to another particle at a distant place [20].

The controlled-NOT (CNOT) gates are useful to make an entangled state. With a CNOT gate, the following operation is performed with the target bit depending on the Control bit.

Control bit 0 Target bit $\downarrow\to\downarrow$, $\uparrow\to\uparrow$
Control bit 1 Target bit $\downarrow\to\uparrow$, $\uparrow\to\downarrow$.

As the initial state, we assume the state of Control bit $C_0\langle 0\rangle + C_1\langle 1\rangle$ and target bit $\langle\downarrow\rangle$. The total state after the operation of the CNOT gate is $C_0\langle 0\downarrow\rangle + C_1\langle 1\uparrow\rangle$, which is an entangled state.

Here we introduce an example of a method to get a CNOT gate with an atom (ion) trapped by a harmonic potential (motion frequency ν_{vib}) with the vibrational energy of $E_V = (n_{vib} + 1/2)h\nu_{vib}$ (Chapter 3.8.4). The vibrational quantum number $n_{vib} = 0, 1$ is the Control bit and the atomic internal energy state (↑ and ↓) is the target bit as shown below.

An electromagnetic wave $P_c\cos\left(2\pi\nu_{\uparrow\downarrow}t\right)$ ($\nu_{\uparrow\downarrow}$:↑↔↓ transition frequency) is irradiated to induce the following transition.

$$(n_{vib},\uparrow) \rightarrow \frac{(n_{vib},\uparrow) + (n_{vib},\downarrow)}{\sqrt{2}}, \quad (n_{vib},\downarrow) \rightarrow \frac{(n_{vib},\uparrow) + (n_{vib},\downarrow)}{\sqrt{2}}$$

An electromagnetic wave $P_{aex}\cos\left\{2\pi\left(\nu_{\uparrow aex} - \nu_{vib}\right)t\right\}$ ($\nu_{\uparrow aex}$: ↑↔ *aex* transition frequency) is irradiated to induce the following transition (aex: another atomic internal state).

$$(n_{vib},\uparrow) \rightarrow (n_{vib} - 1, aex) \rightarrow -(n_{vib},\uparrow)$$

This phase inversion is not caused for $n_{vib} = 0$.

An electromagnetic wave $P_c\cos\left(2\pi\nu_{\uparrow\downarrow}t + \pi\right)$ is irradiated with the period and intensity same as (1).

After the procedure shown above, the following transition is caused.

$$(0,\uparrow) \rightarrow \frac{(0,\uparrow) + (0,\downarrow)}{\sqrt{2}} \rightarrow \frac{(0,\uparrow) + (0,\downarrow)}{\sqrt{2}} \rightarrow (0,\uparrow)$$

$$(0,\downarrow) \rightarrow \frac{(0,\uparrow) + (0,\downarrow)}{\sqrt{2}} \rightarrow \frac{(0,\uparrow) + (0,\downarrow)}{\sqrt{2}} \rightarrow (0,\downarrow)$$

$$(1,\uparrow) \rightarrow \frac{(1,\uparrow) + (1,\downarrow)}{\sqrt{2}} \rightarrow \frac{(1,\downarrow) - (1,\uparrow)}{\sqrt{2}} \rightarrow (1,\downarrow)$$

$$(1,\downarrow) \rightarrow \frac{(1,\uparrow) + (1,\downarrow)}{\sqrt{2}} \rightarrow \frac{(1,\downarrow) - (1,\uparrow)}{\sqrt{2}} \rightarrow (1,\uparrow)$$

7.6 Interaction of single trapped ions with photons in cavities

In a cavity, the number of states of light is quite different from that in a free space. For matters (atoms, molecules, ions) in a cavity, the spontaneous emission rate can be increased or decreased by tuning the cavity length [21]. Black body radiation (BBR) from a finite-size particle has the selectivity of the frequency mode from its size because it has a role as a cavity [22]. The research field about the interaction between a matter and light in a cavity is called "cavity quantum electrodynamics (CQED)."

It is useful to observe the interaction between a single ion trapped in a cavity and a single photon. For an ion with two energy states α and β, the coupling state $\Phi = \cos(\varphi/2)\left(\alpha, n_p = 1\right) + \sin(\varphi/2)\left(\beta, n_p = 0\right)$ can be constructed (n_p: photon number). Keller *et al.* generated continuous single photons using an ion-trap

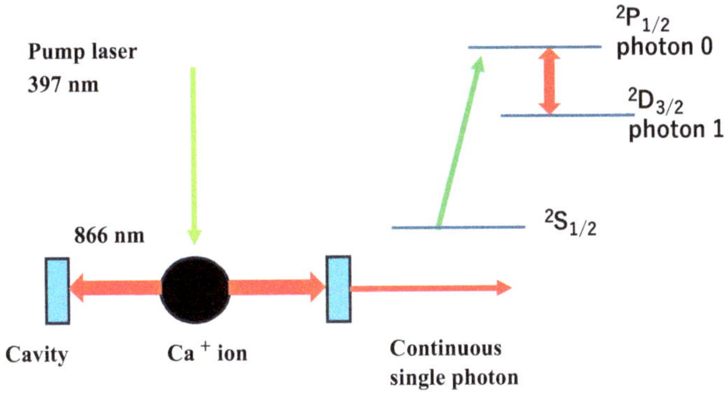

Figure 7.5 Single photon generator using a single trapped $^{40}Ca^{+}$ ion

cavity system [23]. A single $^{40}Ca^{+}$ ion is trapped inside an optical cavity (length 8 mm), which is resonant with the $^2D_{3/2}$ (α) – $^2P_{1/2}$ (β) transition frequency (866 nm). Irradiating a pump laser, resonant to the $^2S_{1/2}$ – $^2P_{1/2}$ transition, a cavity-assisted Raman transition is induced, and a single photon with the wavelength of 866 nm is supplied to the cavity, as shown in Figure 7.5.

Observing the photon from the cavity, the state is determined to be the α ($^2D_{3/2}$) state. On the other hand, the observation of the fluorescence with the wavelength of the pump laser (397 nm), the state is determined to be the β ($^2P_{1/2}$) state. Observing the photon from the cavity and the fluorescence simultaneously, the fluorescence is suppressed when the photon from the cavity is observed. Observing the photon transformed via optical fiber from the cavity, we can determine the state of the trapped ion also at distant places.

7.7 Productions of cold molecules

The research on ultra-low temperature has been performed mainly with laser-cooled atoms. Laser cooling has been performed with atoms with simple energy structures, alkali or alkali-earth atoms. Molecular laser cooling has been believed to be difficult because of complicated energy structures with vibrational and rotational structures. However, production of cold molecules is useful, for example, for precision measurement of molecular transition frequencies. Here, several methods to produce cold molecules are introduced.

7.7.1 Construction of molecules from laser-cooled atoms

We introduce the binding of laser-cooled atoms by photoassociation. The transition frequency of atom A is shifted by the collision with atom B. When there is a bound state between atom A in the excited state and atom B in the ground state, atom A colliding with atom B is excited by a laser light with a frequency lower than the transition frequency (no excitation without collision). If the kinetic energy of relative motion between two atoms is low enough, the combination of atoms A and B is transformed to the electronically excited state of a molecule AB and de-excited to the molecular electronically ground state by spontaneous emission. The energy difference between absorbed and emitted photons is compensated for by the loss of the kinetic energy of the interatomic relative motion, and atoms are bounded as shown in Figure 7.6.

The production of Cs_2 molecules was reported in 1998 [24]. KRb molecules were also produced by photoassociation [25]. AB molecules produced in this method are mostly in a high vibrational-rotational state. However, the transform to the vibrational-rotational ground state was attained using a method called "stimulated Raman adiabatic passage (STIRAP)" [26].

Cold molecules can also be produced by bonding cold atoms through Feshbach resonance. Atoms A and B are in close proximity with ultra-low kinetic energy and high density. Feshbach resonance is the resonance between two states Φ_{atoms} (energy E_a) and Φ_{mol} (energy E_m): Φ_{atoms} is the state of atoms A and B in the ground, and Φ_{mol} is the state where atom A in the higher energy state is bonded with atom B [27]. Applying the magnetic field, E_a and E_m are Zeeman-shifted and both states are mixed. This mixture is significant, particularly when $E_a = E_m$ (bonding energy = energy gap between two states in atom A). Changing the magnetic field adiabatically, there is a repulsion of the crossing of both energy levels, and the transformation between the atomic and molecular states is possible. With this method, all atoms can be transformed into a molecular state as shown in Figure 7.7. Kinetic energies of molecules produced by Feshbach resonance are generally lower than those produced by photoassociation. After Rb_2 [28] and Cs_2 [29] molecules were produced with this method, fermionic $^{40}K^{87}Rb$ molecules [30] were produced with this method. Produced molecules are in a high vibrational-rotational state at this stage. Using STIRAP, produced molecules are transformed to the absolute ground state. By localizing all fermionic $^{40}K^{87}Rb$ molecules into a

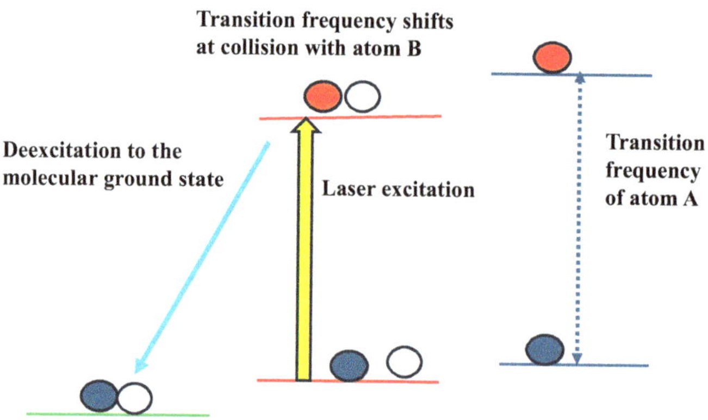

Figure 7.6 The procedure of making cold molecule AB by photoassociation of cold atoms A and B

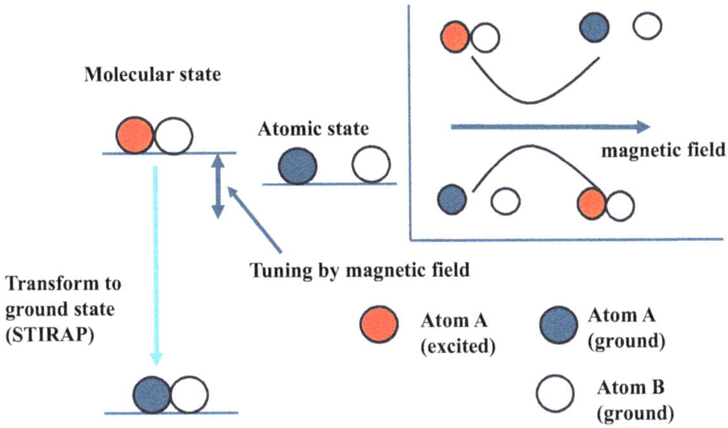

Figure 7.7 *Feshbach resonance between the atomic state and molecular state. Both energies can be tuned using a magnetic field, and there is a repulsion upon the crossing of energy levels. Changing the magnetic field adiabatically, we can transform between the atomic and molecular states. Black and red squares indicate the ground and excited states of atom A, respectively.*

single quantum state (including hyperfine state), the chemical reaction induced by the inelastic collision was suppressed (Chapter 7.3) [31]. The BEC state was attained with NaCs polar molecules with a large electric dipole moment (2.5 D) with a temperature of 300 nK and a density of 10^{12} /cm^3 [32].

7.7.2 Molecular deceleration

7.7.2.1 Buffer-gas cooling

The simplest method to cool atoms or molecules will be cooling via collision with the vapor of liquid helium. With this method, all kinds of atoms and molecules can be cooled down to the order of 1 K. If we can prepare a trap system with a depth of 1 K, we can trap cooled atoms or molecules. The first trapping of cold molecules was reported in 1998 [33]. CaH molecules, produced by the laser ablation of CaH_2 crystal, were cooled by He buffer gas and loaded into a magnetic trap.

7.7.2.2 Laser cooling of molecules

As shown in Chapter 4.2.4, continuous repetition of cycling transition (excitation by laser light + spontaneous emission de-excitation) is required for laser cooling. Laser cooling has been performed with atoms with simple energy structures, and it was believed to be difficult to apply to molecules having vibrational-rotational states. However, it is applicable for some molecules with relatively simple energy structures.

With the electric dipole transition, the rotational quantum number N of diatomic molecules can change only by $\Delta N = \pm 1$. The spontaneous emission transition from $N = 0$ state is possible only to the $N = 1$ state in the lower energy states. Here, we consider the possibility of laser cooling using the $X(n_v, N) = (0, 1) \leftrightarrow A(0, 0)$ transition, where X and A denote the electronic ground and the first excited states, respectively and n_v is the vibrational quantum number. Laser cooling is possible when the spontaneous emission transition is possible only between the same vibrational states. This requirement is satisfied for some molecules, as shown below. The wavefunction of the vibrational states in the A state $\Phi_A(n'_v)$ is given by

$$\Phi_A(n'_v) = \sum c(n'_v, n_v)\Phi_X(n_v) \tag{7.18}$$

and the ratio of the spontaneous emission rate from the $A\ n'_v$ state to the $X\ n_v$ state is proportional to $|c(n'_v, n_v)|^2$ (Franck-Condon factor). When the bond length and the vibrational frequencies on the X and A states are almost equal (Figure 7.8),

$$\begin{aligned} &\Phi_A(n_v) \approx \Phi_X(n_v) \\ &c(n'_v, n_v) \approx 1 \quad (n'_v = n_v) \\ &c(n'_v, n_v) \approx 0 \quad (n'_v \neq n_v). \end{aligned} \tag{7.19}$$

This relation is called “diagonal,” and the spontaneous emission transition is dominated by those between the same vibrational states. Then, the $X(n_v, N) = (0, 1) \leftrightarrow A(0, 0)$ transition repeats about 100 cycles. Using also the repump laser for the $X(n_v, N) = (1, 1) \leftrightarrow A(0, 0)$ and $X(n_v, N) = (2, 1) \leftrightarrow A(1, 0)$ transitions (Figure 7.8), laser cooling of SrF [34] and CaF [35,36] molecules was

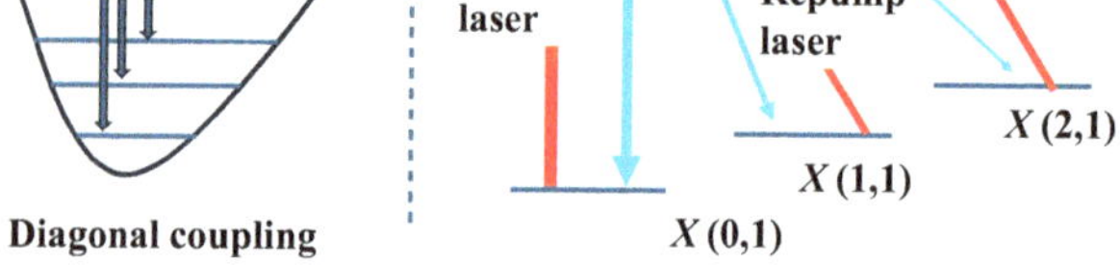

Figure 7.8 Laser cooling is possible for molecules with diagonal coupling, where the bond length in the ground and excited states is almost equal. Laser cooling is performed with a cooling laser for $X(n_v, N) = (0, 1) \leftrightarrow A(0, 0)$ transition and repump lasers for $X(1, 1) \leftrightarrow A(0, 0)$ and $X(2, 1) \leftrightarrow A(1, 0)$ transitions.

achieved. As it is still difficult to repeat the cooing transition more than 10,000 cycles, it is performed after precooling with buffer gas (a few K). Laser cooling was expected to be difficult for polyatomic molecules, but it succeeded with the SrOH molecule (750 μK) [37].

In principle, laser cooling is also possible for CaH, MgH, and SrH molecules. However, these molecules are mostly localized to the $(n_v,N) = (0,0)$ state after the buffer gas precooling, and the population in the $N = 1$ state is very low. Laser cooling is also possible with the $X(n_v, N) = (0,0), (0,2) \leftrightarrow A(0,1)$ cycling transition, but a more complicated laser system is required. To my knowledge, there has been no report of using this transition for laser cooling.

Laser cooling seems to be difficult for molecules, for which the coupling between X and A states is not diagonal.

7.7.2.3 Optoelectrical Sisyphus cooling

Recently, a novel method was developed to reduce the molecular kinetic energy through interaction with a DC electric field, which is applicable for polyatomic polar molecules. Molecules in the electric low-field seeking states (positive energy shift with electric field) move between places with high and low electric fields. The energy shift induced by the electric field depends on the molecular rotational state, and we consider two states with the strong and weak energy shifts: |S> and |W>. At the low electric field region, all molecules are localized to the |S> state by optical pumping (Chapter 4.2.2). When molecules in the |S> state move to the high

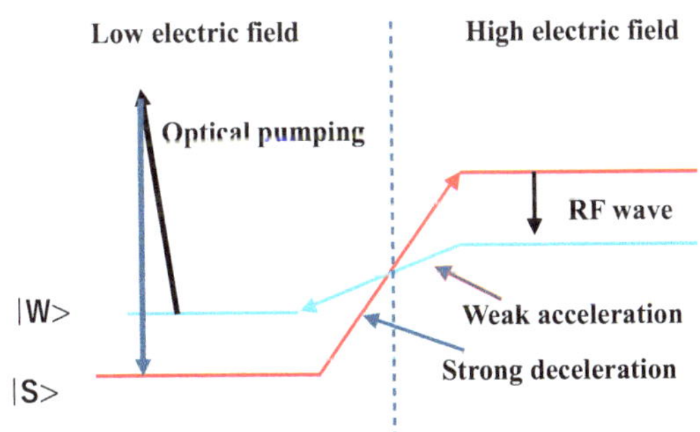

Figure 7.9 The procedure of optoelectrical Sisyphus cooling: molecular cooling by the motion between the high and low electric field regions. In the low electric field region, molecules are pumped to the state with a strong Stark energy shift and move to the high electric field region, getting a strong deceleration. In the high electric field region, molecules are transformed to a state with a weak Stark energy shift by an RF wave. Molecules are accelerated when they move to the low electric field region. Molecules are cooled because the deceleration in a state of the strong Stark shift is larger than the acceleration in the state of the weak Stark shift.

electric field region, molecules are decelerated by the increased potential energy induced by the higher electric field. Then molecules are transformed to the |W> state using an RF wave. Molecules in the |W> move to the low electric field region with an acceleration. With this cycle, the molecular kinetic energy is decreased because the deceleration in the |S> state is larger than the acceleration in the |W> state (Figure 7.9) [38]. The kinetic energy of 420 μK and the single rotational state with the purity of 80% were obtained with the H_2CO molecule [39].

References

[1] Kapitza P. 'Viscosity of liquid helium below the λ-point'. *Nature*, 1938; 141 74. DOI:10.1038/141074a0

[2] Allen J. F. and Misener A. D. 'Flow phenomena in liquid helium II'. *Nature*, 1938; 142 643–644. DOI:10.1038/142643a0

[3] Vinen W. F. 'Vortex lines in liquid helium II'. *Physica*, 1958; 24 513–517

[4] Rayfield G. and Raif F. 'Quantized vortex rings in superfluid helium'. *Phys. Rev.*, 1964;. 136 A1194. DOI:10.1103/PhysRev.136.A1194

[5] Ouboter R. B. 'Heike Kamerlingh Onnes's discovery of superconductivity'. *Sci. Am.*, 1997; 276 98–103

[6] Meissner W. and Ochsenfeld R. 'Ein neuer Effekt bei Eintritt der Supraleitfähigkeit'. *Naturwissenschaft*, 1933; 21 787–788. DOI:10.1007/BF01504252

[7] Snider E., Dasenbrock-Gammon N., McBride R. *et al.* 'Room-temperature superconductivity in a carbonaceous sulfur hydride'. *Nature*, 2020; 586 373–377 DOI: 10.1038/s41586-020-2801-z

[8] Shimizu F., Shimizu K. and Takuma H. 'Double-slit interference with ultracold metastable neon atoms'. *Phys. Rev. A*, 1992; 46 R17–R20. DOI:10.1103/PhysRevA.46.R17

[9] Shimizu F., Shimizu K., and Takuma H. 'Stark phase shift and deflection in the ultracold atomic interferometer'. *Jpn. J. Appl. Phys.*, 1992; 31 L436–L438. DOI:10.1143/JJAP.31.L436

[10] Peter A. Chung K. Y. and Chu S. 'Measurement of gravitational acceleration by dropping atoms'. *Nature*, 1999; 400 849–852. DOI:10.1038/23655

[11] Anderson P. W. 'Pressure broadening in the microwave and infra-red regions'. *Phys. Rev.*, 1949; 76 647–661. DOI:10.1103/PhysRev.76.647

[12] Katori H., Kunigita H. and Ido T. 'Quantum statistical effect on ionizing collisions of ultracold metastable Kr isotopes'. *Phys. Rev. A*, 1995; 52 R4324. DOI:10.1103/PhysRevA.52.R4324

[13] Kajita M. 'Cold collisions between boson or fermion molecules'. *Phys. Rev. A*, 2004; 69 012709. DOI:10.1103/PhysRevA.69.012709

[14] Anderson M. H., Ensher J. R., Matthews M. R., Wieman C. E. and Cornell E. A. 'Observation of Bose-Einstein condensation in a dilute atomic vapor'. *Science*, 1995; 269 198–201. DOI:10.1126/science.269.5221.198

[15] Stamper-Kurn D. M., Andrews M. R., Chikkatur A. P. *et al.* 'Optical confinement of a Bose-Einstein condensate'. *Phys. Rev. Lett.*, 1998; 80 2027–2030. DOI:10.1103/PhysRevLett.80.2027

[16] Hau L. V., Harris S. E., Dutton Z. and Behroozi C. H. ‘Light speed reduction to 17 metres per second in an ultracold atomic gas’. *Nature*, 1999; 397 594–598. DOI:10.1038/17561

[17] Donley E. A., Claussen N. R., Cornish S. L., Roberts J. L., Cornell E. A. and Wieman C. E. ‘Dynamics of collapsing and exploding Bose–Einstein condensates’. *Nature*, 2001; 412 295–299. DOI:10.1038/35085500

[18] Hara H., Konishi H., Nakajima S., Takasu Y. and Takahashi Y. ‘A three-dimensional optical lattice of ytterbium and lithium atomic gas mixture’. *J. Phys. Soc. Jpn.*, 2014; 83 014003. DOI:10.7566/JPSJ.83.014003

[19] Jördens R., Strohmaier N., Günter K., Moritz H. and Esslinger T. ‘A Mott insulator of fermionic atoms in an optical lattice’. *Nature*, 2008; 455 204–207. DOI:10.1038/nature07244

[20] Takeda S., Mizuta T., Fuwa M., van Loock P. and Furusawa A. ‘Deterministic quantum teleportation of photonic quantum bits by a hybrid technique’. *Nature*, 2013; 500 315–318. DOI:10.1038/nature12301

[21] Yuan Y., Qian C., Sun S. *et al.* ‘Enhanced spontaneous emission rate and luminescence intensity of $CsPbBr_3$ quantum dots using a high-Q microdisk cavit’. *J. Phys. Chem. Lett.*, 2025; 16 1095–1102 DOI: 10.1021/acs.jpclett.4c03615

[22] Odashima H., Tachikawa M. and Takehiro K. ‘Mode-selective thermal radiation from a microparticle’. *Phys. Rev. A*, 2009; 80 041806 (R). DOI:10.1103/PhysRevA.80.041806

[23] Keller M., Lange B., Hayasaka K., Lange W. and Walther H. ‘Continuous generation of single photons with controlled waveform in an ion-trap cavity system’. *Nature*, 2004; 431 1075–1078. DOI:10.1038/nature02961

[24] Fioretti A., Comparat D., Crubellier A., Dulieu O., Masnou-Seeuws F. and Pillet P. ‘Formation of cold Cs_2 molecules through photoassociation’. *Phys. Rev. Lett.*, 1998; 80 4402–4405. DOI:10.1103/PhysRevLett.80.4402

[25] Takekoshi T., Patterson B. M. and Knize R. J. ‘Observation of optically trapped cold cesium molecules’. *Phys. Rev. Lett.*, 1998; 81 5105–6108. DOI:10.1103/PhysRevLett.81.5105

[26] Aikawa K., Akamatsu D., Hayashi M. *et al.* ‘Coherent transfer of photo-associated molecules into the rovibrational ground state’. *Phys. Rev. Lett.*, 2010; 105 203001. DOI:10.1103/PhysRevLett.105.203001

[27] Kohler T., Goral K. and Julienne P. ‘Production of cold molecules via magnetically tunable Feshbach resonances’. *Rev. Modern Phys.*, 2006; 78 1311–1361. DOI:10.1103/RevModPhys.78.1311

[28] Lang F., Winkler K., Strauss C., Grimm R. and Hecker Denschlag J. ‘Ultracold triplet molecules in the rovibrational ground state’. *Phys. Rev. Lett.*, 2008; 101 133005. DOI:10.1103/PhysRevLett.101.133005

[29] Danz J. G., Haller E. and Gustavsson M. ‘Quantum gas of deeply bound ground state molecules’. *Science*, 2008; 321 1062–1066. DOI:10.1126/science.1159909

[30] Ni K.-K. ‘A high phase-space-density gas of polar molecules’. *Science*, 2008; 322 231–235. DOI:10.1126/science.1163861

[31] Ospelkaus S., Ni K.-K., Wang D. *et al.* 'Quantum-state controlled chemical reactions of ultracold potassium-rubidium molecules'. *Science*, 2010; 327 853–857. DOI:10.1126/science.1184121
[32] Stevenson I., Lam A. Z., Bigagli N. *et al.* 'Ultracold gas of dipolar NaCs ground state molecules'. *Phys. Rev. Lett.*, 2023; 130 113002. DOI:10.1103/PhysRevLett.130.113002
[33] Weinstein J. D., deCarvalho R., Guillet T., Friedrich B. and Doyle J. M. 'Magnetic trapping of calcium monohydride molecules at millikelvin temperatures'. *Nature*, 1998; 395 148–150. DOI:10.1038/25965
[34] Barry J. F., McCarron D. J., Norrgard E. B., Steinecker M. H. and DeMille D. 'Magneto-optical trapping of a diatomic molecule'. *Nature*, 2014; 512 286–289. DOI:10.1038/25949
[35] Zhelyazkova V., Cournol A., Wall T. E. *et al.* 'Laser cooling and slowing of CaF molecules'. *Phys. Rev. A*, 2014; 89 053416. DOI:10.1103/PhysRevA.89.053416
[36] Anderegg L., Augenbraun B. L., Chae E. *et al.* 'Radio frequency magneto-optical trapping of CaF with high density'. *Phys. Rev. Lett.*, 2017; 119 103201. DOI:10.1103/PhysRevLett.119.103201
[37] Kozyryev I., Baum L., Matsuda K. *et al.* 'Sisyphus laser cooling of a polyatomic molecule'. *Phys. Rev. Lett.*, 2017; 118 173201. DOI:10.1103/PhysRevLett.118.173201
[38] Zeppenfeld M., Motsch M., Pinkse P. W. H. and Rempe G. 'Optoelectrical cooling of polar molecules'. *Phys. Rev. A*, 2009; 80 041401(R). DOI:10.1103/PhysRevA.80.041401
[39] Prehn A., Ibrügger M., Glöckner R., Rempe G. and Zeppenfeld M. 'Optoelectrical cooling of polar molecules to submillikelvin temperatures'. *Phys. Rev. Lett.*, 2016; 116 063005. DOI:10.1103/PhysRevLett.116.063005

Chapter 8

Particles and antiparticles

Abstract

For all particles, there are also antiparticles with conjugate electric charge. The existence of the positron (antiparticle of the electron) was predicted by the Dirac equation. Particles and antiparticles are produced and annihilated as a pair. The characteristics of antiparticles are almost the same as those of particles except for the conjugate electric charge, but there must be some slight difference to explain the particle-dominant universe. The unequal abundance of particles and antiparticles is explained by the violation of the CP symmetry (charge conjugate + mirror image), which has been discovered with K^0 and B^0 mesons. The Kobayashi-Maskawa theory derived the possible violation of CP symmetry. On the other hand, the CPT symmetry (charge conjugate + mirror image + time reversal) has been confirmed from the equalities of transition frequencies of hydrogen and antihydrogen atoms.

8.1 Antiparticles predicted by the Dirac equation

The Dirac equation is a formula to derive energy eigenvalues and eigen wavefunctions of an electron (mass of m_e) at each state (Chapter 3.9.2). Each state gives four eigenfunctions with spin states $M_S = \pm 1/2$ with a rest energy of $\pm m_e c^2$. The existence of negative rest energy was a mystery. Dirac proposed a model in which the vacuum is a sea of electrons with negative rest energy, as shown in Figure 8.1 [1]. There is a finite number of quantum states with negative rest energy in the vacuum, and all states are normally filled with electrons. The electron is a fermion (Chapter 7.2), and multiple electrons cannot exist in a single quantum state; therefore, electrons with positive energy cannot decay to the negative energy state. Giving energy of $2m_e c^2$ to an electron in the vacuum, transition to the state of the positive rest energy is induced, and an electron having positive rest energy and negative electric charge is produced. The lack of an electron in the vacuum (electron sea) is observed as a particle with the positive electronic charge and the mass of m_e (positron). This phenomenon is called "pair production." When an electron decays from a positive rest energy state to a negative rest energy state, the electron and positron annihilate simultaneously, and the energy of $2m_e c^2$ is produced, which

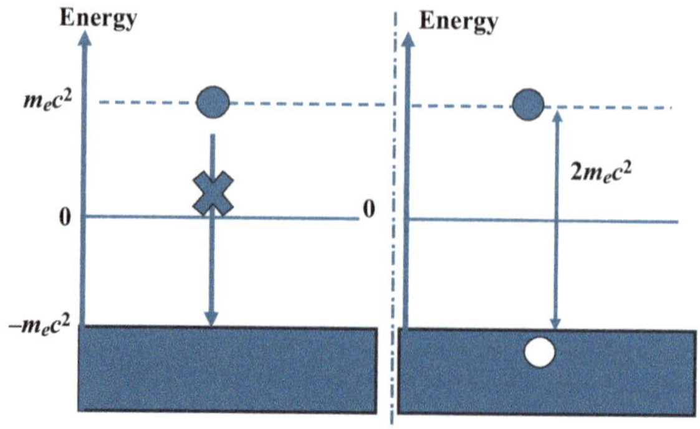

Figure 8.1 *Electron sea model proposed by Dirac. The vacuum is filled with electrons with negative rest energy. An electron with positive rest energy cannot decay to negative rest energy when all states in negative rest energy are filled. Giving energy of $2m_ec^2$, the electron is excited to positive rest energy. In the vacuum, a lack of electrons with negative rest energy is recognized as the appearance of a particle with mass of m_e and a positive electric charge.*

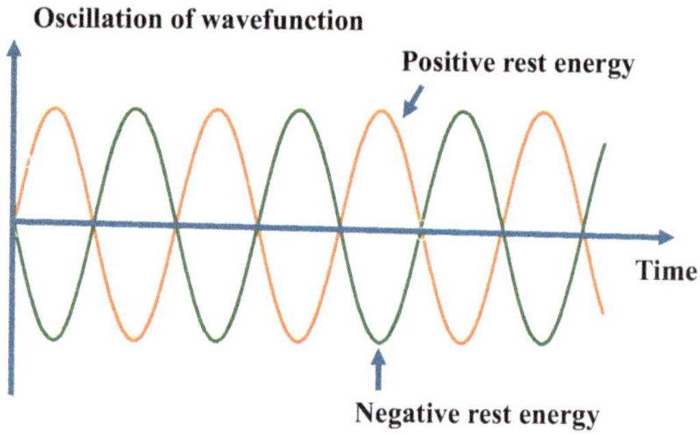

Figure 8.2 *Oscillation of wavefunction with positive rest energy (particle) and negative rest energy (antiparticle)*

is called "pair annihilation." This prediction was confirmed by the discovery of the positron in 1932 [2], which was the first discovery of an antiparticle. Pair production and pair annihilation of electron and positron were also observed. Antiprotons (negative charge) and antineutrons were also discovered in 1955 and 1956, respectively. It is now confirmed that there are antiparticles for all particles. However, the idea of an electron sea is not used for boson particles (no limited number in a quantum state), because all boson particles should decay to the negative rest energy state in the vacuum. A new interpretation was provided by Feynman [3].

The difference between positive and negative rest energy is interpreted as the inverse oscillation direction of the wavefunction, as shown in Figure 8.2. It is

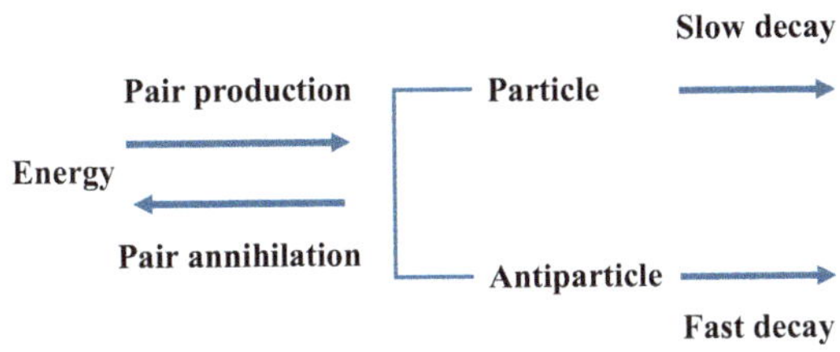

Figure 8.3 Pair production and annihilation of particle and antiparticle. The slightly larger number of particles may be explained by the difference in decay rates.

reasonable to interpret that oscillations in both directions equally exist. An electromagnetic field gives energy shifts to particles, which is interpreted as a change in the oscillation frequency. As the antiparticle has a conjugate electric charge, the change of the frequency is caused in the inverted direction, and the absolute values of frequencies of the particle and antiparticle are the same, also with the electromagnetic field.

Pair production and pair annihilation are caused by all kinds of particles, as shown in Figure 8.3. If the production and annihilation of particles and antiparticles were caused only by these procedures, the numbers of particles and antiparticles must be equal.

To describe the relation between particle and antiparticle, the following symmetries were proposed.

Charge (C) symmetry: All images of charge conjugation of real phenomena are equally possible, as the real phenomena

Parity (P) symmetry: All mirror images of real phenomena are equally possible with real phenomena

Time (T) symmetry: All images of time reversal of real phenomena are equally possible as real phenomena

If these symmetries are strictly held, the number of particles and antiparticles must be exactly equal. However, we do not see antiparticles in the universe. With the current model, both particles and antiparticles existed for several seconds after the birth of the universe. Antiparticles were annihilated via collision with particles. The number of particles exceeded that of antiparticles with the ratio of ($10^9 + 1$: 10^9), and particles remained after all antiparticles were annihilated. There must be some violation of these symmetries, for example, by the slight difference in decay time as shown in Figure 8.3.

8.2 Early research on radioactivity

To discuss the unequal abundance of particles and antiparticles, it would be important to summarize the radioactive decays, in which numbers of particles and antiparticles change without pair production or annihilation. At first, the X-ray (an electromagnetic wave with a frequency higher than the ultraviolet area) was discovered by Röntgen in 1895 [4]. In 1896, Becequerel discovered a natural radioactive ray from uranium. Its identity was clarified by Rutherford to be α-ray ($^4He^{2+}$: two protons and two neutrons), β-ray (electron), and γ-ray (electromagnetic wave with frequency higher than X-ray). The γ-ray was clarified not to be the flow of charged particles because its path was not bent by the magnetic field.

The radioactive rays are emitted with a change in the structure of the nucleus. The nuclei can take only discrete energy eigenvalues because protons and neutrons are bound in an area five orders smaller than the atomic size. Therefore, radioactive rays should have discrete energies. The energy of the α-ray is distributed in a narrow range. However, the energy of electrons emitted as the β-ray is distributed continuously in a wide range. Pauli gave a hypothesis that not only electrons but also an unknown particle (named neutrino) is simultaneously emitted at the β-decay. The sum of the energy of the electron and the neutrino is discrete [5]. In 1932, the neutron was discovered, and the β-decay was clarified to be a transition from neutron n into (proton p + electron e^- + neutrino ν), which is possible because the mass (rest energy) of the neutron is slightly larger than that of the proton.

Neutrinos are electrically neutral. The mass of the neutrino was assumed to be zero at first. In 1956, neutrinos were discovered by observing the production of a neutron and a positron via the reaction between a neutrino and a nucleus in a water molecule [6]. There are three kinds of neutrinos: electron-neutrino (ν_e), μ-neutrino (ν_μ), and τ-neutrino (ν_τ). There are also antineutrinos: $\overline{\nu_e}$, $\overline{\nu_\mu}$, and $\overline{\nu_\tau}$. The spin is 1/2, but only $M_S = -1/2$ state exists with a neutrino, and only $M_S = 1/2$ state exists with an antineutrino. Therefore, C symmetry is violated with the neutrino because the spin state is different. After distinguishing three kinds of neutrino and antineutrino, the β-decay is clarified to be $n \rightarrow p + e^- + \overline{\nu_e}$.

The direction of emitted electrons at the β-decay of Co atoms is localized in the direction antiparallel to the nuclear spin, as shown in Figure 8.4 [7]. This result shows that P symmetry is also violated at the β-decay. Currently, all physical phenomena are described by four interactions: strong interaction, electromagnetic interaction, weak interaction, and gravitational interaction. The β-decay is induced

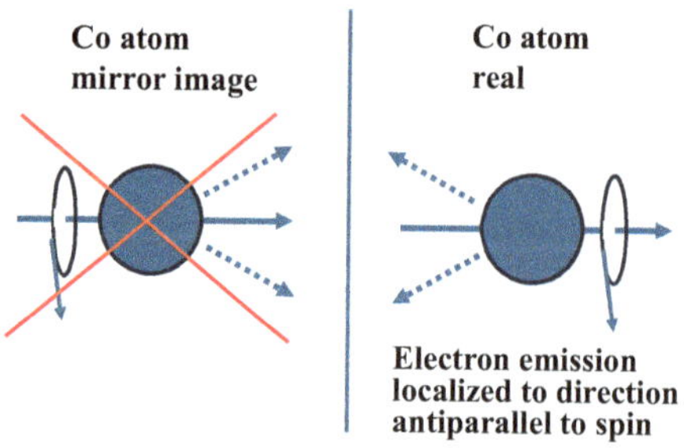

Figure 8.4 The emission of an electron by the β-decay is localized in the inverse direction to the spin

by the weak interaction, which violates the P symmetry and plays a significant role in making an inhomogeneous abundance of particles and antiparticles.

8.3 Violation of CP symmetry

As shown in Chapter 8.2, the C and P symmetries are violated with weak interaction. However, CP symmetry (charge conjugate + mirror reflection) was expected to hold, because $\overline{\nu_e}$ is the mirror image of ν_e. Here we compare the decay of muon μ^- and its antiparticle μ^+ (Chapter 8.4). The CP symmetry between the $\mu^- \rightarrow e^- + \overline{\nu_e} + \nu_\mu$ and $\mu^+ \rightarrow e^+ + \nu_e + \overline{\nu_\mu}$ transitions hold, although the C and P symmetries are violated for the direction distributions of emitted particles.

The violation of the CP symmetry was discovered by observing the two kinds of mixtures of particles and antiparticles of the K^0-meson (antiparticle $\overline{K^0}$). Chapter 9.4 indicates the identity of K^0 and $\overline{K^0}$. The lifetime of K^0_S $(= [K^0 + \overline{K^0}]/\sqrt{2})$ is 9.0×10^{-11} s, which is much shorter than that for K^0_L $(= [K^0 - \overline{K^0}]/\sqrt{2}$ lifetime 1.2×10^{-8} s). Therefore, pure K^0_L should be attained after the lifetime of K^0_S. But K^0_S was observed also after a much longer period than its lifetime, as shown in Figure 8.5 [8]. This result indicates that the $K^0_L \rightarrow K^0_S$ transition (change of symmetry) exists. The change of symmetry of an electrically neutral particle indicates the violation of the CP symmetry. The violation of the T symmetry was also indicated from experimental results showing that the $K^0 \rightarrow \overline{K^0}$ transition rate is lower than the $\overline{K^0} \rightarrow K^0$ transition rate with a ratio of 0.66% [9].

The difference of lifetime between B_0 and $\overline{B_0}$ with the order of 10% was experimentally confirmed [10]. This result indicates the violation of the T symmetry. The inequality of the numbers of particles and antiparticles is derived from the difference of the lifetime, as shown in Figure 8.3.

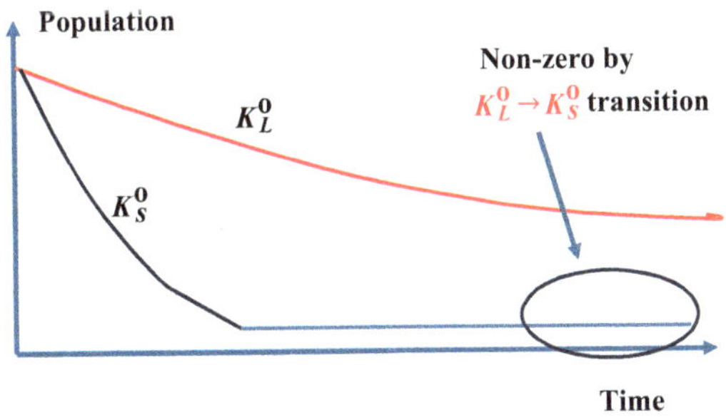

Figure 8.5 *Decay of populations of two kinds of K^0-mesons: K^0_L and K^0_S. The lifetime of K^0_S is very short, but its population is not zero also after the lifetime because of the $K^0_L \rightarrow K^0_S$ transition.*

8.4 Three generations of particles and CP violation

There are μ-particle (μ^-) and τ-particle (τ^-), which have almost the same properties as the electron (e^-) except for the larger mass and shorter lifetime. The masses of μ^- and τ^- are $250m_e$ and $3550m_e$ (m_e: electron mass 9.1×10^{-31} kg, 5.51 MeV), respectively. Both particles decay with

$$\mu^- \rightarrow e^- + \overline{\nu_e} + \nu_\mu \text{ lifetime } 2.2 \times 10^{-6} \text{ s}$$

$$\tau^- \rightarrow \pi^- + \pi^0 + \nu_{\tau,} \quad 25.5\ \%$$

$$\rightarrow e^- + \overline{\nu_e} + \nu_{\tau,} \quad 17.8\ \%$$

$$\rightarrow \mu^- + \overline{\nu_\mu} + \nu_{\tau,} \quad 17.4\ \% \text{ lifetime } 2.9 \times 10^{-13} \text{ s.}$$

We consider μ^- and τ^- as the second and third generations of electrons. Particles in the higher generation decay and transform to the lower generation. There are also antiparticles of e^-, μ^-, and τ^-; e^+, μ^+, and τ^+.

The electron-neutrino (ν_e), μ-neutrino (ν_μ), and τ-neutrino (ν_τ) are the neutrinos in the first, second, and third generations. The periodic change of the abundance ratio of three neutrinos (neutrino oscillation) was theoretically analyzed [11]. However, the observed neutrino oscillation should not be observed if the mass of neutrinos were zero and they moved with the speed of light, because the theory of relativity prohibits any temporal change in a frame moving with the speed of light. Neutrino oscillation was observed by different research groups, and this result indicates that the masses of neutrinos are nonzero. There is currently a project to compare neutrino oscillation between ν and $\overline{\nu}$, which might indicate the violation of the CP and T symmetries [12]. Previous experimental results indicate a slight difference between the abundance ratio of ν_e and $\overline{\nu_e}$, but this difference is not large enough in comparison with current measurement uncertainty. A higher sensitivity of the neutrino signal is required to confirm with the lower statistical measurement uncertainty.

The three generations of elementary particles give a theoretical explanation of the violation of CP symmetry. The first generation particle p_1 is in real, not pure p_1, but coupling between three generations $p_{real} = c_1p_1 + c_2p_2 + c_3p_3$. The wavefunction of antiparticles are complex conjugates of that of particles and $\overline{p_{real}} = c_1^*\overline{p_1} + c_2^*\overline{p_2} + c_3^*\overline{p_3}$. For the coupling between two states, the coefficients are transformed to real values, and there is no difference between particle and antiparticle. With coupling between three states, the coefficients cannot be transformed into real values, and there is a phase difference between particle and antiparticle. Kobayashi and Maskawa explained the violation of CP symmetry with quarks, predicting three generations of quarks before quarks in the third generation were discovered (detailed in Chapter 9.4) [13].

8.5 Search for the Electron Electric Dipole Moment (eEDM)

Kobayashi-Maskawa theory gave the difference in numbers of particles and antiparticles, which is ten orders smaller than the result obtained from the ratio of particles and photons. There may be some other mechanism to induce the violation of CP symmetry. Several research groups are eager to search for the electric dipole moment (EDM) in the electron (eEDM). If nonzero eEDM exists, there are at least two states $\pm d_e$ with distinct directions (Figure 8.6). The electron has two spin states, $M_s = \pm 1/2$. If the electron spin and eEDM are independent states, more than four electrons could exist in one (n, L, M_L) state in an atom (n: principal quantum number, L: orbital angular momentum quantum number, M_L: magnetic quantum number). The possible eEDM state can be one state corresponding to the spin state: the possible two-electron states are ($M_s = 1/2, +d_e$) and ($M_s = -1/2, -d_e$). Time reversal gives transformations ($M_s = 1/2, +d_e$) $\rightarrow$ ($M_s = -1/2, +d_e$) and ($M_s = -1/2, -d_e$) $\rightarrow$ ($M_s = 1/2, -d_e$), which are not possible; therefore, T symmetry is violated. The P symmetry is also violated as the mirror image conversion gives the transformation ($M_s = 1/2, +d_e$) $\rightarrow$ ($M_s = 1/2, -d_e$). If the measurement of eEDM is much larger than the value estimated by the Kobayashi-Maskawa theory (10^{-38} e cm) [13], it might give a consistent estimation of the abundance ratio between particles and antiparticles.

The search for the eEDM should be done by measuring the energy shift induced by the electric field E_e. Atoms or molecules having nonzero electron spin are used for the measurement, because free electrons are accelerated by an electric field. When the magnetic and electric fields are zero, the ($M_s = 1/2, +d_e$) and ($M_s = -1/2, -d_e$) states are degenerated. A magnetic field is applied to distinguish the direction of the electron spin and eEDM. Given a magnetic field B_m, there are two Zeeman energy shifts with coefficients $+a_Z M_S$ and the transition frequency between both states is

$$\nu_Z = 2\frac{a_Z B_m}{h}. \tag{8.1}$$

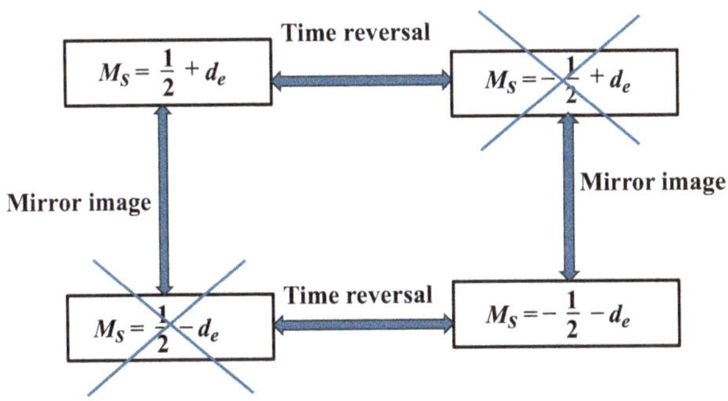

Figure 8.6 Existence of the electron electric dipole moment (eEDM) as evidence for the violation of T and P symmetries

The coefficient a_Z is given by the relation between electron spin and the electron orbital angular momentum (without electron orbital angular momentum $a_Z/h = 1.4$ MHz/G). When an electric field E_e is also applied parallel or anti-parallel to the magnetic field, an additional energy shift is induced (called Stark shift), and the transition frequency $\nu_\pm$ is given by

$$\nu_\pm = \nu_Z \pm \frac{2d_e E_e}{h}. \tag{8.2}$$

Measuring the difference between these two transition frequencies,

$$\nu_+ - \nu_- = \frac{4d_e E_e}{h} \tag{8.3}$$

yields a value for the eEDM.

Values of eEDM were estimated to be 10^{-26}–10^{-30} e cm by several theorists. With these estimated values, the Stark shift might be detectable when an electric field of 10^{10}–10^{11} V/cm is applied. Such high electric fields may be experienced by electrons moving with high velocity in atoms or molecules, because the relativistic effect is significant and the electromagnetic field in the electron frame is different from that in the laboratory frame. Such electrons experience a high magnetic field and a high electric field simultaneously. These electrons are not accelerated because forces from both fields are balanced. The direction of the internal electric field is random, but it can be aligned by applying an external electric field. The electric field applied to electron is several orders higher than applied external electric field.

For the search for eEDM, a high internal electric field and a precision frequency measurement are required. The internal electric field is high with atoms or molecules having a heavy nucleus. To align the direction of the internal electric field, polar molecules are advantageous. For the precision measurement of transition frequency, it is preferable to use cold atoms or molecules because we can observe the spectrum with a narrow line width, taking a long interaction time between the microwave and the atoms or molecules. The quadratic Doppler effect is also suppressed by using cold atoms or molecules. Cold polar molecules having heavy nuclei are most advantageous.

In 2011, a group at Imperial College London obtained an upper limit for eEDM of 1.05×10^{-27} e cm using YbF molecules (electron spin $\pm 1/2$, electron orbital angular momentum 0) in a thermal beam [14]. In 2014, a joint group from Harvard and Yale obtained an upper limit for eEDM of 8.7×10^{-29} e cm from a measurement using ThO molecules cooled by the vapor of liquid He (Chapter (7.7.2.1)) [15]. The measurement uncertainty was reduced drastically by using cooled molecules. The Zeeman coefficient in the ThO transition frequencies (3 kHz/G) is much smaller than in the YbF molecule (1.4 MHz/G), because the Zeeman shifts induced by electron spin (± 1) and electron angular momentum (∓ 2) cancel each other (total angular momentum ± 1). Therefore, the influence of the fluctuation of the magnetic field is three orders smaller than the measurement with the YbF molecule. In 2018, this upper limit was reduced to 1.1×10^{-29} e cm after revising the measurement apparatus [16]. This group is continuing the revision of the apparatus, aiming at the reduction of the measurement uncertainty by one order. A group in the Joint Institute of the University of Colorado Boulder and NIST (JILA) attained the upper limit of 4.1×10^{-30} e cm using trapped HfF^+ molecular ion (Zeeman coefficient 3 kHz/G) [17]. Recently, the group at Imperial College London succeeded in decelerating YbF molecules by laser cooling (Chapter 7.7.2.2) [18]. The search for eEDM might be possible using laser-cooled molecules.

Previously, a nonzero value of eEDM has never been obtained by experiment. The upper limit of the value of eEDM has been reduced to below several theoretical predictions (10^{-26}–10^{-29} e cm). Therefore, the previous experimental results gave important results to discuss the validity of theories to estimate the violation of the T symmetry.

8.6 Tests of CPT symmetry

When the CP symmetry is violated, the T symmetry is also violated, as shown in Chapter 8.3. The CPT symmetry (charge conjugation + mirror reflection + time reversal) is conserved with the decay of K^0 and B^0 mesons.

To guarantee the Lorenz invariance (all physical laws in a coordinate must hold with any coordinates given by the Lorenz transform), the CPT symmetry must be conserved strictly [19]. A simple interpretation of the CPT symmetry is shown using the Dirac equation (valid for particles with spin 1/2)

$$\left(\frac{h}{2\pi i}\frac{\partial}{\partial t} - q_e\Phi_{el}\right) = \sum_{Q=x,y,z}\begin{pmatrix}0 & \sigma_Q\\ \sigma_Q & 0\end{pmatrix}\left(\frac{h}{2\pi i}\frac{\partial}{\partial Q} + q_eA_Q\right) + \begin{pmatrix}I & 0\\ 0 & -I\end{pmatrix}mc^2$$

σ_Q: Pauli matrix $Q = x, y, z$

A_Q: components of the magnetic field vector potential in the Q-direction

Φ_{el}: electricvoltage. (8.4)

Giving the CPT-transform ($q_e \to -q_e$, $Q \to -Q$, $t \to -t$) to (8.4),

$$\left(\frac{h}{2\pi i}\frac{\partial}{\partial t} - q_e\Phi_{el}\right) = \sum_{Q=x,y,z}\begin{pmatrix}0 & \sigma_Q\\ \sigma_Q & 0\end{pmatrix}\left(\frac{h}{2\pi i}\frac{\partial}{\partial Q} + q_eA_Q\right) + \begin{pmatrix}-I & 0\\ 0 & I\end{pmatrix}mc^2 \tag{8.5}$$

is obtained. Equation (8.5) is also valid, transforming the solution of the wave-function into a four-dimensional vector by $\begin{pmatrix}\vec{u}\\ \vec{w}\end{pmatrix} \to \begin{pmatrix}\vec{w}\\ \vec{u}\end{pmatrix}$, where $\vec{u}$ and $\vec{w}$ are two-dimensional vectors.

To conserve the CPT symmetry, equalities of the absolute value of the electric charge and the mass are required between the particles and the antiparticles. The equality of |charge| / mass between proton and antiproton was confirmed by comparing the frequencies of cyclotron motion (circular motion of a charged particle under the magnetic field). The measurement uncertainty was reduced to 1.6×10^{-11} [20].

After the production of anti-H atoms (antiproton + positron), comparisons of transition frequencies between H and anti-H atoms have been performed. The equality of the 1S→2S transition frequencies between the H atom and anti-H atom was confirmed with an uncertainty of 2×10^{-12} [21]. This transition is one-photon forbidden and observed by two-photon absorption of the laser light with a wavelength of 250 nm. Low measurement uncertainty is obtained because the spontaneous emission rate is very low (one-photon forbidden transition), and the spectrum linewidth is narrow. The 1S→2P transition frequencies were also compared, and they were in good agreement within the uncertainty of 5×10^{-8} [22]. The measurement uncertainty of the 1S→2P transition frequency is much higher than the 1S→2S transition frequencies because the 1S→2P transition is one-photon allowed, and the spectrum linewidth given by the spontaneous emission rate is much larger than the 1S→2S transition. However, the development of a laser source (125 nm) to observe the 1S→2P transition made it possible to perform laser cooling of H and anti-H atoms. Laser cooling of an anti-H atom was recently demonstrated [23]; therefore, the further reduction of the measurement uncertainty is expected for the anti-H transition frequency.

A novel method was developed to measure the $^2S_{1/2}$ $F = 0 \rightarrow 1$ hyperfine transition frequency, which is applicable also for antihydrogen [24]. This transition is more sensitive to the violation of the CPT symmetry than the 1S→2S transition frequency; therefore, the confirmation of the equality of the hyperfine transition frequency is useful to confirm the CPT symmetry.

8.7 Gravitational interaction of matter and antimatter

There was a question whether the gravitational force between matter and antimatter is attractive or repulsive.

Following the equivalence principle between the inertial force and the gravitational force, the gravitational force between antimatter and matter seems to be the same as that between matter. Observing an object from an accelerating frame, we see this object accelerating in the opposite direction, regardless of matter or antimatter. The inertial force is common for matter and antimatter, as shown in Figure 8.7. If the equivalence principle is valid also for antimatter, the gravitational force should be common also for the antimatter.

But from the CPT symmetry, the idea of the repulsive force between matter and antimatter seemed to be more advantageous, while we interpret the gravitational force as the attractive force between matter. The electromagnetic force on antimatter works in the inverse direction because of the conjugate electric charge.

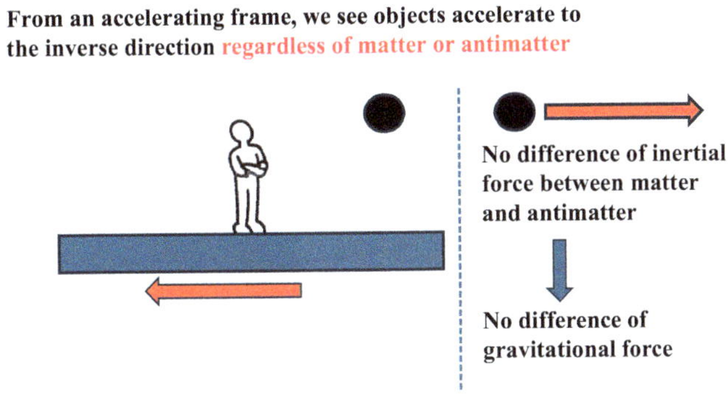

Figure 8.7 The equality of the inertial force for matter and antimatter

With this motion, the CPT symmetry holds between matter and antimatter. If the gravitational force works in the same direction for matter and antimatter, CPT symmetry is violated.

A recent experiment conducted at the European Organization for Nuclear Research (CERN) in Switzerland suggested that matter and antimatter particles responded to gravity in the same way, with an accuracy of 97% [25]. Is the CPT symmetry violated? Interpreting the motion in the gravitational field as the inertial motion in the distorted space, the motion of matter and antimatter is the same, and this result might be explained without the violation of CPT symmetry.

8.8 Are neutrinos Majorana particles?

The wavefunction of antiparticles is the complex conjugate of that of the particles. Majorana suggested in 1937 that electrically neutral particles with the spin of 1/2 can be described by a real-valued wave equation, and then wavefunctions of particles and antiparticles are equal. Such particles (called Majorana particles) are their own antiparticles [26]. There is a hypothesis that the neutrino is a Majorana particle, which may be confirmed by the discovery of the neutrino-less double β-decay shown in Figure 8.8. [27]; two neutrons in the same nuclear decay ($n \rightarrow p + e^- + \overline{\nu_e}$) simultaneously. If the neutrino is a Majorana particle, two neutrinos can be pair annihilated, and it is observed as a neutrino-less double β-decay. The emitted energy at the pair annihilation gives information about the neutrino mass, which has never been measured.

The two-neutrino double β-decays have been observed with 14 isotopes, including ^{48}Ca, ^{76}Ge, and ^{82}Se. Their lifetimes are of the order of 10^{21} years. However, the neutrino-less double β-decay has never been discovered. Its lifetime has been experimentally estimated to be longer than 10^{25} years [28]. It seems difficult to detect neutrino-less double β-decay separated from two-neutrino double β-decays, whose rate is more than four orders higher.

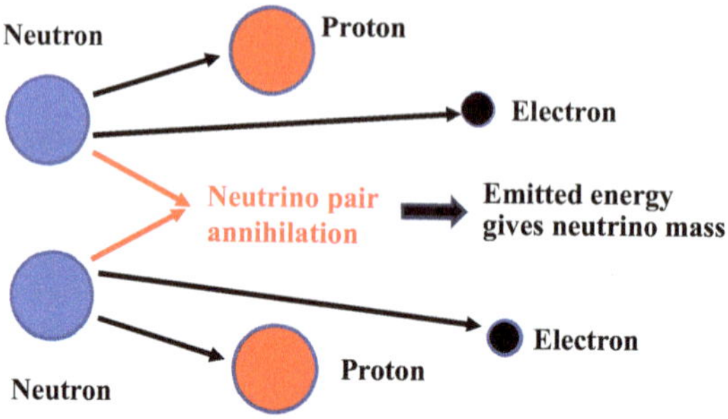

Figure 8.8 *When double β-decay occurs, two protons, electrons, and neutrinos are produced. If the neutrino is a Majorana particle, pair annihilation can be caused. The discovery of the neutrino-less double β-decay proves that the neutrino is a Majorana particle.*

References

[1] Dirac P. A. M. A. 'A theory of electrons and protons'. *Proc. R. Soc. Lond. A*, 1930; 126 360–365. DOI:10.1098/rspa.1930.0013

[2] Hanson N. R. *The Concept of the Positron*, Cambridge University Press, 1963; 136–139.

[3] Feynman R. P. 'Space-time approach to non-relativistic quantum mechanics'. *Rev. Mod. Phys.*, 1948; 20 367 1–36, DOI:10.1103/RevModPhys.20.367

[4] Röntgen W. *Ueber eine neue Art von Strahlen. Verlaeufige Mitteilung. Sitzungsberichten der Wuerzburger Physik-Medic Gesellschaft Wuerzburg*, 1895; 137

[5] Webpage from RDIOACTIVITY.EU.COM (The neutrino hypotheisi) https://radioactivity.eu.com/phenomenon/neutrino_hypothesis

[6] Wintr K. *Neutrino Physics. Cambridge U. Press.*, 2000; p 33.

[7] Wu C. S., Ambler E., Hayward R. W., Hoppes D. D. and Hudson R. P. 'Experimentaltest of parity conservation in beta decay'. *Phys. Rev.*, 1957; 105 1413–1414. DOI:10.1103/PhysRev.105.1413

[8] Christenson J. H., Cronin J. W., Fitch V. L. and Turlay R. 'Evidence for the 2π decay of the K_2^0 Meson'. *Phys. Rev. Lett.*, 1964; 13 138–140. DOI:10.1103/PhysRevLett.13.138

[9] Angelopoulos A., Apostolakis A., Aslanides E. *et al.* 'First direct observation of time-reversal non-invariance in the neutral-kaon system'. *Phys. Lett. B*, 1998; 444 43–51. DOI:10.1016/S0370-2693(98)01356-2

[10] Schwarzschild B. M. 'Time-reversal asymmetry in particle physics has finally been clearly seen'. *Phys. Today*, 2012; DOI:10.1063/PT.3.1774

[11] Maki Z., Nakagawa M. and Sakata S. 'Remarks on the unified model of elementary particles'. *Prog. Theor. Phys.*, 1962; 28 870–880. DOI:10.1143/PTP.28.870

[12] Georgescu L. 'CP violation and beyond'. *Nat. Rev. Phys.*, 2020; 2 120–122. DOI:10.1038/s42254-020-0159-7

[13] Kobayashsi M. and Maskawa T. 'CP-violation in the renormalizable theory of weak interaction'. *Prog. Theor. Phys.*, 1973: 49(2) 652–657. DOI:10.1143/PTP.49.652

[14] Hudson J. J., Kara D. M., Smallman I. J., Sauer B. E., Tarbutt M. R. and Hinds E. A. 'Improved measurement of the shape of the electron'. *Nature*, 2011; 473 493–496. DOI:10.1038/nature10104

[15] Baron J., Campbell W. C., DeMille D. *et al.* 'Order of magnitude smaller limit on the electric dipole moment of the electron'. *Science*, 2014; 343 269–272. DOI:10.1126/science.1248213

[16] Andreev V., Ang D. G., DeMille D. *et al.* 'Improved limit on the electric dipole moment of the electron'. *Nature*, 2018; 562 355–360. DOI:10.1038/s41586-018-0599-8

[17] Roussy T. S., Caldwell L., Wright T. *et al.* 'An improved bound on the electron's electric dipole moment'. *Science*, 2023; 381 46–50. DOI:10.1126/science.adg4084

[18] Lim J., Almond J. R., Trigatzis M. A. *et al.* 'Laser cooled YbF molecules for measuring the electron's electric dipole moment'. *Phys. Rev. Lett.*, 2018; 120 123201 DOI:10.1103/PhysRevLett.120.123201

[19] Tureanu A. 'CPT and lorentz invariance: their relation and violation'. *J. Phys. Conf. Ser.*, 2013; 474 012031. DOI:10.1088/1742-6596/474/1/012031

[20] Borchert M. J., Devlin J. A., Erlewein S. R. *et al.* 'A 16-parts-per-trillion measurement of the antiproton-to-proton charge-mass ratio'. *Nature*, 2022; 601 53–57. DOI:10.1038/s41586-021-04203-w

[21] Ahmadi M. 'Characterization of the 1S–2S transition in antihydrogen'. *Nature*, 2018; 557 71–75. DOI:10.1038/s41586-018-0017-2

[22] Ahmadi M. 'Observation of the 1S–2P Lyman-α transition in antihydrogen'. *Nature*, 2018; 561 211–215 DOI.10.1038/s41586-018-0435-1

[23] Baker C. J., Bertsche W., Capra A. *et al.* 'Laser cooling of antihydrogen atoms'. *Nature*, 2021; 592 35–42. DOI:10.1038/s41586-021-03289-6

[24] Nowak L., Malbrunot C., Simon M. C. *et al.* 'CPT and Lorentz symmetry tests with hydrogen using a novel in-beam hyperfine spectroscopy method applicable to antihydrogen experiments'. *Phys. Lett. B*, 2024; 858 139012. DOI:10.1016/j.physletb.2024.139012

[25] Anderson E. K. 'Observation of the effect of gravity on the motion of antimatter'. *Nature*, 2023; 621 716–722. DOI:10.1038/s41586-023-06527-1

[26] Majorana E. 'Teoria simmetrica dell'elettrone e del positrone'. *Il Nuovo Cimento*, 1937; 14 171–184. DOI:10.1007/BF02961314

[27] Rodejohann W. 'Neutrino-less double beta decay and particle physics'. *Int. J. Mod. Phys. E*, 2011; 20 1833–1930. DOI:10.1142/S0218301311020186

[28] Aalseth C. E., Algrall N., Aguayo E. *et al.* 'Search for neutrinoless double-β decay in ^{76}Ge with the Majorana Demonstrator'. *Phys. Rev. Lett.*, 2018; 120132502. DOI:10.1103/PhysRevLett.120.132502

Chapter 9

High-energy physics and the structure of matter

Abstract

This chapter introduces the structure of nuclei, which can be searched by the collision of particles with velocities close to the speed of light. Giving ultra-high collision energy, protons and neutrons were clarified to be bound by three quarks. The mass of protons and neutrons are two orders larger than the total mass of constructing quarks, because the kinetic energies of quarks confined inside the proton radius are very large (uncertainty principle between position and momentum). The large kinetic energies of quarks make the pair production of quarks and antiquarks in the proton or neutron possible, which induces the nuclear force. The properties of strong interaction and weak interaction are introduced. The violation of CP symmetry was made by the phase transition from a high-energy state to a low-energy state. In the high-energy state, the system is symmetric, and all particles move at the speed of light. With this situation, we cannot distinguish the difference between electromagnetic fields and weak interactions (unified interaction called electroweak interaction). In the low-energy state, Higgs particles interact with quarks, leptons, and weak bosons. Because of the large mass of weak bosons, weak interaction works only over short distances, violating CP symmetry. Grand Unification Theory was proposed to unify strong interaction and electroweak interaction; however, this theory has not been established because predicted phenomena (for example, proton decay) have never been discovered.

9.1 High-energy collisions and pair production

Chapter 8.1 indicates pair production of particles and antiparticles. The pair production of an electron and positron is possible just by irradiating γ-ray to a target. In the vacuum, pair production and annihilation of electron and positron repeat by the fluctuation of vacuum energy, which is the foundation of quantum electrodynamics (QED). Higher energy is required to produce particles with larger masses. Here we consider the pair production of particles A and anti-particles $\overline{A}$ (mass m_A) by the collision between two particles B (mass m_B) with the velocity of $\pm v_x$. The energy of two particles B is given by (Chapter 3.6)

$$E_B = 2\sqrt{(m_B c^2)^2 + (cp_x)^2} > 2cp_x$$
$$p_x = \frac{m_B v_x}{\sqrt{1 - \left(\frac{v_x}{c}\right)^2}}. \tag{9.1}$$

The required velocity v_x is roughly given by

$$2m_A c^2 < 2cp_x = \frac{2m_B v_x c}{\sqrt{1-\left(\frac{v_x}{c}\right)^2}}$$
$$m_A\sqrt{1 - \left(\frac{v_x}{c}\right)^2} < m_B\left(\frac{v_x}{c}\right)$$
$$\frac{v_x}{c} > \frac{m_A}{\sqrt{m_A^2 + m_B^2}}. \tag{9.2}$$

When $m_A = m_B$, the collisional velocity is required to be larger than $c/\sqrt{2}$. Synchrotron accelerator with large radius r_0 is required to minimize the deceleration by irradiation of an electromagnetic wave, which is inversely proportional to r_0^2 (Chapter 3.3.4). The experimental research of elementary particles and nuclei has been conducted with the development of an accelerator, with which particles are accelerated close to the speed of light.

9.2 The nuclear force

The electromagnetic interaction is the exchange of photons. Coulomb's law indicates that the electric field from a point electric charge is proportional to r^{-2}, where r is the distance from the point electric charge. This phenomenon is interpreted as the electric field being proportional to the density of the electric field line from the electric charge, which is inversely proportional to the surface area of the sphere surrounding the electric charge. The electric potential energy is proportional to r^{-1}.

There is another force to bind protons and neutrons in nuclei, called "nuclear force." Nuclear force is also caused by the exchange of a particle X. Considering that X has some finite decay rate, the potential energy is given by (called Yukawa potential)

$$P_Y(r) \propto \frac{\exp\left(-\frac{r}{\lambda_Y}\right)}{r}$$
$$\lambda_Y = v_X \tau'_X$$
$$\tau'_X = \frac{\tau_X}{\sqrt{1 - (v_X/c)^2}}, \tag{9.3}$$

where v_X and τ_X denote the mean velocity and decay time of X, respectively. The decay time in the nucleus should be considered with τ'_X taking the relativistic effect with the motion of X into account. For example, $1/\lambda_Y = 0$ when the mass of X is zero because it can move only at speed of light ($v_X = c$). Using the relativistic definition of momentum $p_x = mv_x/\sqrt{1 - \left(\frac{v_X}{c}\right)^2}$ and the de Broglie wavelength $\lambda_{dB}(= h/p_X)$, the following relation is derived.

$$\lambda_Y = \frac{v_x \tau_x}{\sqrt{1 - (v_X/c)^2}} = \frac{p_X \tau_X}{m} = \frac{h \tau_X}{m \lambda_{dB}}. \tag{9.4}$$

Assuming $\lambda_{dB}/\tau_X = c$ ($\tau_X > \lambda_{dB}/c$ is required for the decay of whole wave broadening), λ_Y corresponds to the Compton wavelength estimated by $mc^2 = cp_X$. $\lambda_Y(= h/mc)$ is predicted from the size of the nucleus. Yukawa predicted the existence of a particle with the mass of the order of $0.11 m_p$, where m_p is the mass of proton (1.6×10^{-27} kg, 938 MeV/c^2 $m_p/m_e = 1840$) [1]. Three kinds of π-mesons with the electric charge of $\pm e$ and neutral (π^+, π^-, and π^0) were discovered afterward. π^- is an antiparticle of π^+. The antiparticle of π^0 is itself.

The mass and lifetime of π-mesons are

$\pi^\pm$ 0.149 m_p 2.6×10^{-8} s
π^0 0.144 m_p 8.5×10^{-17} s.

Physical phenomena are described by four fundamental forces (strong interaction, electromagnetic interaction, weak interaction (induce β-decay), and gravitational interaction). While proton was believed to be an elemental particle, the nuclear force was treated as "strong force." After the discovery of quarks, strong interaction is treated as the force to binds quarks (Chapter 9.3). Nuclear force is a force that comes from the strong force between quarks.

9.3 Structure of the proton and neutron

Protons and neutrons were believed to be elemental particles that cannot be divided further. However, there was a mystery that neutrons have a nonzero magnetic dipole moment despite a neutral electric charge. In 1969, the proton was clarified to be a bound of three particles from the experimental results with high-energy collisions with electrons [2]. High-energy collision was required to discover the structure of proton, because the energy gap between different states is very large for particles bound in an area five orders smaller than electron orbits in atoms (Chapter 3.8.2).

There is an up quark (u) with the electric charge of $+2e/3$ and down quark (d) with the electric charge of $-e/3$. Proton is constructed by two up quarks and one down quark, while a neutron is constructed by one up quark and two down quarks. There are antiparticles for both quarks; $\overline{u}$ and $\overline{d}$. The masses of up and down quarks are on the order of 0.002 m_p and 0.005 m_p, respectively. The force to bind quarks is "strong interaction," which is given by exchanging eight kinds of gluons. Gluons are boson particles (spin 1) without mass or electric charge. The β-decay was interpreted as $d \rightarrow u + e^- + \overline{\nu_e}$.

The mass of a proton and a neutron is two orders larger than the total mass of constructing quarks, which is explained as follows. From the uncertainty principle between position and momentum, the momentums of quarks are roughly estimated by

$$p_{qu} = \frac{h}{4\pi r_p}, \tag{9.5}$$

where r_p is the proton radius (8.3×10^{-16} m). The rest energy of proton is sum of energies of three quarks given by (9.1), which is given by

$$m_p c^2 > 3cp_{qu}, \tag{9.6}$$

and $m_p > 6 \times 10^{-28}$ kg is derived. The proton mass is $m_p = 1.6 \times 10^{-27}$ kg and this rough estimation is reasonable. Therefore, the proton mass comes from the high kinetic energy of quarks, from which pair production and annihilation of quark + antiquark repeat at a high rate. From pair produced quarks, π-mesons are constructed as follows

$$\pi^+ = (u\overline{d})\ \pi^- = (\overline{u}d)\ \pi^0 = \frac{(u\overline{u}) - (d\overline{d})}{\sqrt{2}}.$$

The lifetime of π^0 is much shorter than $\pi^\pm$ because of pair annihilation between quarks and antiquarks.

The strong interaction is described by the interaction between color charges in analogy with the three original colors of light. Quarks have color charges of red, blue, and green. Quarks tend to bind to each other so that the total mixture of colors becomes white, just like electric charge becomes neutral by the bunch of positive and negative charges. The color charges of protons and neutrons are white by the mixture of three original colors as shown in Figure 9.1. Antiquarks have charges of complementary colors. The total color of π-meson can be white by the bunching of one quark and one antiquark. The existence of Ω_{sss} (binding of sss, where s is a strange quark shown in Chapter 9.4) is proof of three kinds of color charges, because quarks are fermions and the binding of three quarks is possible only when they are in three different states.

Previously, quarks have been discovered only in the bunched state with the total color of white because the strong interaction between quarks becomes stronger with a larger distance. The electric charge is observed only as an integer multiple of the elementary charge.

It is a mystery how the spin of a proton or neutron is determined, although it has been clarified that the contribution of quarks is less than 30% and the role of spin and orbital angular momentum of gluons is rather significant.

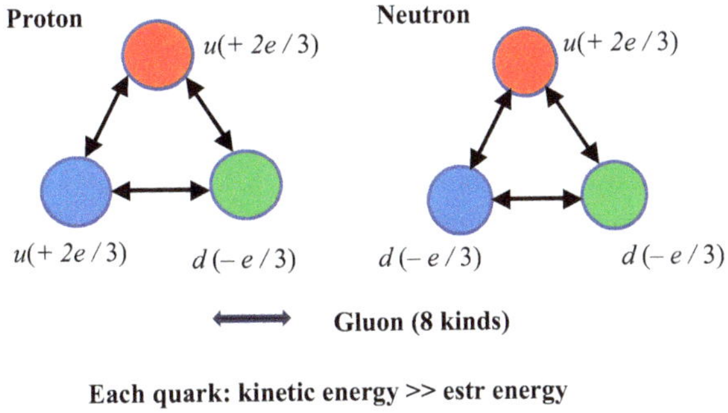

Figure 9.1 *Proton and neutron are constructed by up (u: electric charge +2e/3) and down (d: electric charge −e/3) quarks. The binding is caused by three color charges (red, blue, and green), and the mixture color becomes white. The binding force (strong interaction) is caused by the exchange of eight kinds of gluons. The kinetic energy of a quark is much higher than its rest energy, which makes the mass of a proton and neutron much larger than the total mass of quarks.*

No phenomena have been observed indicating that electrons and neutrinos are bound of smaller particles. Therefore, they are elemental particles called "lepton."

Here, we also introduce weak interaction. The "weak interaction" is several orders weaker than the strong interaction and electromagnetic interaction. It is induced by the exchange of W^+ (electric charge $+e$), W^-(electric charge $-e$), or Z (electric neutral) boson particles. W^- boson is the antiparticle of W^+ boson. Antiparticle of the Z boson is itself. The weak interaction works in a very small area (one order smaller than the proton radius 8.4×10^{-16} m) because the boson particles have large mass (W boson 80 m_p and Z boson 90 m_p) and the motion range before decay (lifetime is on the order of 10^{-9}–10^{-6} s) is very small. W and Z particles were experimentally discovered in 1983 [3]. The weak interaction has very special characteristics because it can violate the C- and P-symmetries. It can induce the transition from the down quark to the up quark. The β-decay is interpreted with the following two procedures:

$$d \rightarrow u + W^-$$
$$W^- \rightarrow e^- + \overline{\nu_e}.$$

9.4 Higher-generation quarks

As shown in Chapter 8.4, μ^- and τ^- are the second and third generations of electrons having the same properties as electrons except for a larger mass and shorter lifetime. The electron-neutrino (ν_e), μ-neutrino (ν_μ), and τ-neutrino (ν_τ) are the neutrinos in the first, second, and third generations. There are three generations also for quarks as shown in Table 9.1. Quarks have never been discovered as single isolated particles, and their existence was recognized by discovery of different mesons. Some examples of mesons are listed in Table 9.2. Here, K^0_S and K^0_L are symmetrical and antisymmetrical coupling between K^0 $(d\bar{s})$ and $\overline{K^0}$ $(\bar{d}s)$.

The transition between quarks induced by the weak interaction was described well, treating quarks as a mixture of different generations. This mixture is induced by strong interaction. In 1963, Cabibbo proposed the Cabibbo angle θ_C with which the β-decay is given by the $d' \to u$ transition, where [4]

$$\langle d\rangle' = \cos\theta_C \langle d\rangle + \sin\theta_C \langle s\rangle. \tag{9.7}$$

Taking $\theta_C = 0.0724\pi$, the experimental result was described well. After the discovery of charm quark (c), the $s' \to c$ transition was also observed with

$$\langle s\rangle' = -\sin\theta_C \langle d\rangle + \cos\theta_C \langle s\rangle. \tag{9.8}$$

From these relations, the Cabibbo matrix was proposed to give the mixture between two generations, causing the $d' \to u$ and $s' \to c$ transitions:

$$\begin{pmatrix} \langle d\rangle' \\ \langle s\rangle' \end{pmatrix} = \begin{bmatrix} \cos\theta_C & \sin\theta_C \\ -\sin\theta_C & \cos\theta_C \end{bmatrix} \begin{pmatrix} \langle d\rangle \\ \langle s\rangle \end{pmatrix}. \tag{9.9}$$

Table 9.1 List of particles from the first to third generations. Mass is given in the lower row with the unit of proton mass m_p and electron mass m_e. The neutrino mass has not been determined.

	Charge	**Generation 1**	**Generation 2**	**Generation 3**
Quark	$+2e^+/3$	Up quark (u)	Charm quark (c)	Top quark (t)
		$0.002m_p$	$1.3m_p$	$183m_p$
	$-e^-/3$	Down quark (d)	Strange quark (s)	Bottom quark (b)
		$0.005m_p$	$0.11m_p$	$4.5m_p$
Lepton	$-e$	Electron (e^-)	μ-particle (μ^-)	τ-particle (τ^-)
		m_e	$250m_e$	$3550m_e$
	0	Electron-neutrino (ν_e)	μ-neutrino (ν_μ)	τ-neutrino (ν_τ)

Table 9.2 List of mesons, which are discussed in this book

Meson	**Antimeson**	**Structure**	**Mass (m_p)**	**Lifetime (s)**
π^+	π^-	$u\bar{d}$	0.148	2.6×10^{-8}
π^0	π^0	$(u\bar{u} - d\bar{d})/\sqrt{2}$	0.144	8.4×10^{-17}
K^+	K^-	$u\bar{s}$	0.527	1.2×10^{-8}
K^0_S	K^0_S	$(d\bar{s} + \bar{d}s)/\sqrt{2}$	0.531	8.9×10^{-11}
K^0_L	K^0_L	$(d\bar{s} - \bar{d}s)/\sqrt{2}$	0.531	5.2×10^{-8}
B^+	B^-	$u\bar{b}$	5.63	1.7×10^{-12}
B_0	$\overline{B_0}$	$d\bar{b}$	5.63	1.5×10^{-12}

The Cabibbo matrix is described using only one real parameter. Describing the coupling coefficients using complex numbers, eq. (9.7) is rewritten as

$$\langle d \rangle' = \cos\theta_C \langle d \rangle + \exp(i\delta_p)\sin\theta_C \langle s \rangle. \tag{9.10}$$

The effect of $\exp(i\delta_p)$ is interpreted as the phase shift between wavefunctions $\langle d \rangle$ ($\propto \exp(2\pi i \nu_d t)$) and $\langle s \rangle$ $(\propto \exp(2\pi i \nu_s t + i\delta_p))$. Just shifting the origin of time $t \to t' = t + \delta_p/(\nu_d - \nu_s)$, the wavefunctions are given by $\langle d \rangle$ ($\propto \exp(2\pi i \nu_d t')$) and $\langle s \rangle$ ($\propto \exp(2\pi i \nu_s t')$), and the effect of $\exp(i\delta_p)$ is eliminated. While considering two generations of quarks, the violation in the CP symmetry is not derived because the value of θ_C is same also for the $\overline{d'} \to \overline{u}$ and $\overline{s'} \to \overline{c}$ transitions. However, considering the coupling between three states, the effect of the phase shift cannot be eliminated. Kobayashi and Maskawa indicated that the coupling between quarks in the three generations is given by complex numbers [5]. The coupling coefficient is given by the Cabibbo-Kobayashi-Maskawa (CKM) matrix

$$\begin{pmatrix} \langle d \rangle' \\ \langle s \rangle' \\ \langle b \rangle' \end{pmatrix} = \begin{bmatrix} 1-\lambda^2/2 & \lambda & A\lambda^3(\rho - i\eta) \\ -\lambda & 1-\lambda^2/2 & A\lambda^2 \\ A\lambda^3(1-\rho-i\eta) & -A\lambda^2 & 1 \end{bmatrix} \begin{pmatrix} \langle d \rangle \\ \langle s \rangle \\ \langle b \rangle \end{pmatrix}$$
$$\lambda = 0.2257, \quad A = 0.814, \quad \rho = 0.135, \quad \eta = 0.349. \tag{9.11}$$

The wavefunctions for antiparticles are the complex conjugates of those for particles. Therefore, the phases of wavefunctions of particles and antiparticles are inverse.

This theory was not accepted soon after its publication, because only three kinds of quarks (*u*, *d*, and *s*) were discovered at that time. It took 21 years until all other quarks (*c*, *b*, and *t*) were discovered, because very high energy is required to produce them. Kobayashi-Maskawa theory derived the unequal numbers of particles and antiparticles. However, the ratio of the number of particles to that of photons estimated by Kobayashi-Maskawa theory is ten orders smaller than the result of observation.

Hyper nuclei include hyperons, which include higher-generation quarks [6]. Λ-hyperons are bound of *ud* and a quark in higher generations. The precision measurement of binding energy of hypernuclei gives important information about nuclear force. Λ^0 (structure *uds*, mass $1.19m_p$, electrically neutral) decays to $p + \pi^-$ or $n + \pi^0$. The lifetime was theoretically estimated to be the order of 10^{-23} s, but it is much longer (1.6×10^{-10} s). This discrepancy was explained by introducing a quantum number "strangeness." Strangeness (given by the number difference of s and $\overline{s}$) is conserved with the strong interaction and electromagnetic interaction and the decay rate is suppressed. The lifetime of Λ^0 is limited by the weak interaction, which can change the strangeness. Λ-particles slide into the center of the hyper nucleus, and it binds the protons and neutrons more tightly due to its stronger nuclear force than conventional nucleus. For example, the ^{7}Be nucleus is 17% smaller. There is also Σ^+-particle with the structure of *uus* (lifetime 8×10^{-11} s electric charge $+e$). There are also Ξ -particles including two high-generation quarks, [7]. Ξ^0 is the binding of *uss* having zero electric charge and mass of $1.40m_p$. Ξ^- is binding of *dss* having $-e$ electric charge and mass of $1.41m_p$. The lifetime of Ξ^0 and Ξ^- are same order as that of Λ^0, because they are limited by the decay of *s* induced by the weak interaction.

No fourth-generation particles were discovered previously. The abundance ratio between proton and neutron is 6:1 (estimated from the ratio of numbers of H and He atoms), which is consistent with the model with the three generations. However, we cannot perfectly deny the existence of the fourth-generation particles. We can imagine very large mass and very short lifetime with fourth-generation particles, and they are also difficult to discover if they exist.

9.5 Unification of forces and the Higgs field

Current physics is described with four fundamental forces: strong interaction, electromagnetic interaction, weak interaction, and gravitational interaction. Einstein had an inspiration that these interactions are just different views of one interaction. He tried the unification of four interactions, but he could not succeed with it. But his fundamental concept is still supported; he could not succeed because he started with the gravitational interaction, with which the treatment is most difficult.

The violation of P-symmetry by weak interaction was a serious problem in unifying it with other forces. The hypothesis of "spontaneous symmetry breaking" was proposed to solve this mystery [8]. The symmetric state with high energy ends up with a phase transition to an asymmetric state with lower energy. Figure 9.2 describes the energy distribution in *x*-direction, which is symmetric but minimum at $x = \pm x_0$. With high energy, the position can be centered $x = 0$ and the distribution is symmetrical. With low energy, the position is shifted to one of $x = \pm x_0$.

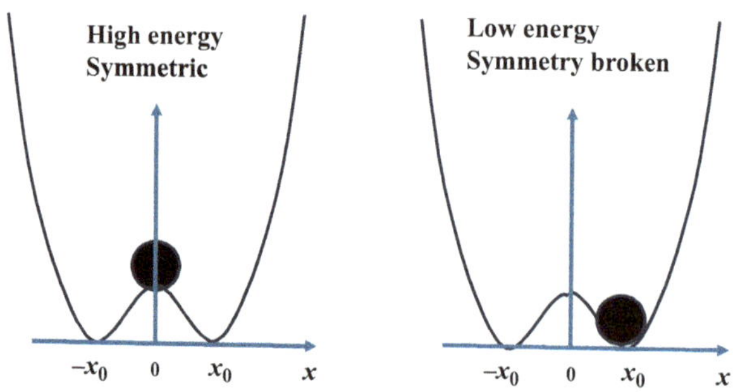

Figure 9.2 The concept of "spontaneous symmetry breaking." The energy distribution for x is symmetric, but it is minimum at $x = \pm x_0$. With high energy, the position can be at $x = 0$, and it is symmetric. With low energy, the position is one of $x = \pm x_0$, and it is no longer symmetric.

Although the probabilities of moving to both positions are equal, the distribution is no longer symmetric after it is shifted to one position.

We can interpret the phase transition to a lower energy state in analogy with the phase transition to the Bose-Einstein condensation (BEC) shown in Chapter 7.4. With high energy, the atomic waves have random phases (symmetric distribution). With low energy, the atoms are transformed to the BEC state, with which the phase is uniform (asymmetric phase distribution).

With the phase transition of $W^{\pm}$ and Z weak bosons, weak interaction became an interaction that violates the P-symmetry. Another hypothesis was also given, indicating that weak bosons have mass with through their interaction with Higgs particles [9]. Higgs particle is a boson particle with zero spin, and the mass is $133m_p$. Quarks and leptons also have mass through their interaction with Higgs particles. Higgs particles evaporate at temperatures higher than 10^{15} K, and all particles move with the speed of light because there is no mass without Higgs particles.

Wineberg and Salam gave a theory to unify the electromagnetic interaction and weak interaction [10]. With high energy, masses of weak bosons are zero and the violation of the P-symmetry is not caused. Therefore, we cannot recognize the difference between weak interaction and electromagnetic interaction. We call the unified interaction "electroweak interaction."

Higgs particle was experimentally discovered in 2012, giving high energy by collision between protons with the velocities close to the speed of light (Figure 9.3). It was difficult to discover Higgs particles because of their large mass. The probability of producing Higgs particles by a single collision between two protons is below 10^{-13}. The direct observation of produced Higgs particles is also difficult because of their ultra-short lifetime ($<10^{-22}$ s). Higgs particles decay to a pair of bottom quarks ($b\bar{b}$), a weak boson pair ($W^{\pm}$), or two photons in the γ-ray region. The properties of Higgs particles were searched for from the energy and momentum produced by the decay of Higgs particles. The existence for Higgs particle was

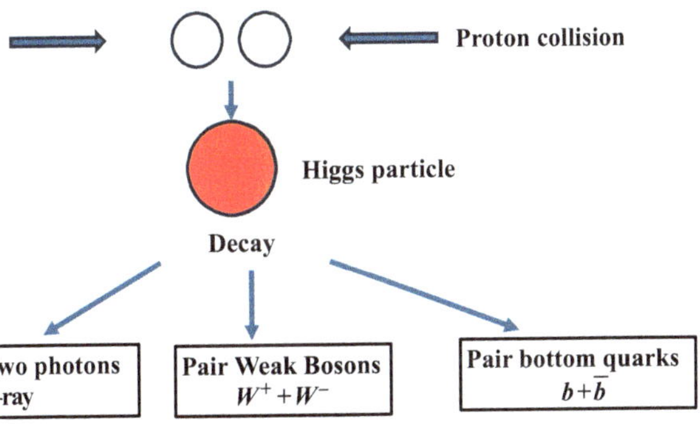

Figure 9.3 The experimental search of Higgs particle from the production after the decay

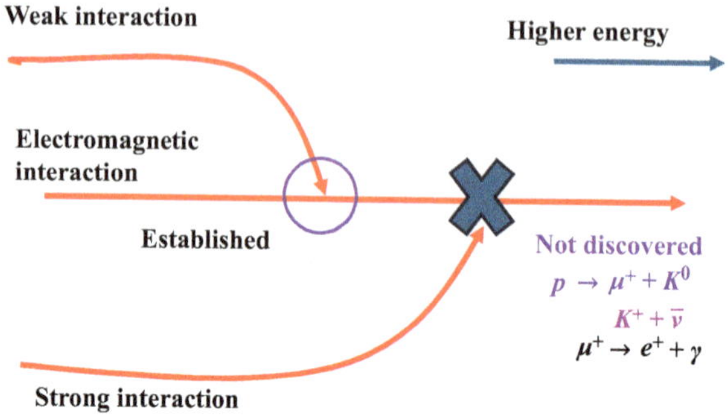

Figure 9.4 The unification of electromagnetic interaction and weak interaction at high energy. Grand Unification Theory insists that strong interaction is also unified with higher energy, but this theory is not established yet because some predicted phenomena have never been discovered.

confirmed from the consistency between the measurements of emitted γ-ray reported by two collaborations [11,12].

A grand unified theory (GUT) is a model in which the strong interaction is also unified with the electroweak interaction at higher energy as shown in Figure 9.4. Quarks have been observed only as bound states until the plasma state was observed with the temperature of 2×10^{12} K. With higher energy (previously not attained), the quarks might be in separate states, and the property of strong interaction might be same as that for the electroweak interaction. GUT gives an explanation that observed electric charge is only an integer multiple of elemental electric charge $\pm e$. However, GUT is not completed yet because phenomena predicted by GUT have never been discovered. Note that free neutrons have the β-decay ($n \rightarrow p + e^- + \overline{\nu_e}$) with a lifetime of 881 s. GUT predicts the possibility of proton decay $p \rightarrow \mu^+ + K^0$ or $\overline{\nu} + K^+$. The mean lifetime of a proton is estimated to be 10^{30}–10^{32} years. The experiment to detect the proton decay has been performed at Super-Kamiokande, having 500 million liters of distilled water at 1000 m underground in Kamioka, Gifu Prefecture in Japan. The decay of a proton from one of 10^{33} water molecules might be observed over several years. The lower limit of mean lifetime was estimated to be 3.6×10^{33} years [13]. Currently, there is also a project to discover the decay of $\mu^+ \rightarrow e^+ + \gamma$, which can be evidence of GUT (not derived from previous standard theory). This decay has previously never been discovered, and the upper limit of the probability was estimated to be 1.5×10^{-13} [14].

9.6 The strong CP-problem

There is a mystery called "strong CP-problem" [15]. With the strong interaction of the color charge, there is a parameter called "vacuum angle," which cannot be determined uniquely. With the nonzero vacuum angle, the CP symmetry is violated. However, the conservation of the CP symmetry has been confirmed experimentally, taking the upper limit of the vacuum angle 10^{-9} radian. It is a mystery why the vacuum angle is so small. To solve this problem, the existence of a spinless particle (axion) with a small mass was predicted. The color charge (strong interaction) produces an effective periodic potential, and the axion has an oscillation around the position where the potential is minimum (zero vacuum angle) [16]. Axion is expected to be in the Bose-Einstein condensate, with which all axions are in the lowest energy state. Axion is one candidate for dark matter, and many researchers are eager to discover it, as shown in Chapter 6.7.3. If a nonzero vacuum angle is discovered, it might explain the unequal abundance of particles and antiparticles.

9.7 Quantum effects in gravitational fields

Currently, the unification of theory of general relativity and quantum mechanics has never been attained. First, we consider the condition that the quantum effect have some effects on the gravitational field.

Matters with any value of mass m are blackhole when their radius is below Schwarzshild radius (Chapter 3.7)

$$R_{SW} = \frac{2Gm}{c^2}, \tag{9.12}$$

where G is the gravitational constant. On the other hand, Compton wavelength λ_{Com} is defined by

$$mc^2 = c|\vec{p}| = \frac{ch}{\lambda_{Com}}$$

$$\lambda_{Com} = \frac{h}{mc}. \tag{9.13}$$

Planck mass and Planck length are defined by

$$R_{SW} = \frac{\lambda_{Com}}{\pi}$$

$$m_{PL} = \sqrt{\frac{ch}{2\pi G}} = 2.17 \times 10^{-8}\ \text{kg}$$

$$l_{PL} = \sqrt{\frac{hG}{2\pi c^3}} = 1.62 \times 10^{-35}\ \text{m} \tag{9.14}$$

Matters with the mass smaller than Planck mass cannot be black holes, because the de Broglie wavelength is larger than the Schwarzshild radius. The Planck length is the minimum Schwarzshift radius. With the size smaller than the Planck length, previous physical laws are not applicable. Planck time is defined as

$$\tau_{PL} = \frac{l_{PL}}{c} = 5.4 \times 10^{-44}\ \text{s} \tag{9.15}$$

and Planck temperature is defined as

$$T_{PL} = \frac{h}{k_B \tau_{PL}} = 1.42 \times 10^{32}\ \text{K}. \tag{9.16}$$

Planck temperature is expected to be the highest temperature at which the current physical laws are valid. The temperature above the Planck temperature has never been artificially attained, but the temperature at the Big Bang might be higher. It is predicted that the phase transition to a lower energy state (temperature of 10^{15} K) occurred 10^{-34} s after the Big Bang.

Superstring theory has been proposed to unify the theory of general relativity and quantum mechanics [17]. Taking particles as points, the propagation time is zero, and the energy becomes infinite from the uncertainty principle of time and energy. With the superstring theory, particles are considered to have a finite size with the order of Planck length. Then the energy cannot be higher than the Planck temperature. The strings have vibrational energy, which gives the rest energy.

The superstring theory requires ten or more dimensions, although we know only four dimensions (positions in three directions and time). Then, what are the extra dimensions? We cannot give a clear explanation. The Big Bang might have induced expansion only in the four dimensions, while the size in extra dimensions remains small. The four-dimensional coordinate axes are lines at macro-scale, but they may be cylinders having an extra dimension at microscale. The distortion of the coordinate axis by gravitational force may cause interaction with extra dimensions, because the coordinates of extra dimensions forming the cylindrical coordinate system are also distorted. The interaction with extra dimensions is not possible for other interactions (strong interaction, electromagnetic interaction, weak interaction). The gravitational interaction is much weaker than other interactions because of the leak of energy to extra dimensions. There might be different combinations of fundamental constants in the world of whole dimensions. We may be seeing only the world in a four-dimensional space with a suitable combination of fundamental constants. We cannot see out of the four-dimensional space, where we cannot exist. We can imagine a train that can move only on a one-dimensional railroad given on a two-dimensional plane.

It seems difficult to give experimental evidence of the superstring theory because Planck length is more than 18 orders smaller than the size of nuclei. However, it might be useful to search for applications of gravitational laws in the microscale (confirmed with distance larger than 1 mm), because the existence of extra dimensions might be confirmed from the violation of universal gravitation at short distances. The detailed search of the gravitational interaction at ultra-short distances is difficult because the interatomic potential with distance below 1 nm is dominated by the electromagnetic interaction. The interaction between neutrons might give us useful information about extra dimensions.

References

[1] Eidelman S., Hayes, K. G., Olive, K. A. *et al.* 'Review of particle physics'. *Phys. Lett. B*, 2004; 592 1–4. DOI:10.1016/j.physletb.2004.06.001

[2] Bloom E. D., Coward D. H., DeStaebler H. *et al.* 'High-energy inelastic $e - p$ scattering at 6° and 10°'. *Phys. Rev. Lett.*, 1969; 23 930–934. DOI:10.1103/PhysRevLett.23.930

[3] Navas S., Amsler C., Gutsche T. *et al.* 'Review of particle physics'. *Phys. Rev. D*, 2024; 110 030001. DOI:10.1103/PhysRevD.110.030001

[4] Cabibbo N. 'Unitary symmetry and leptonic decays'. *Phys. Rev. Lett.*, 1963; 10 531–533. DOI:10.1103/PhysRevLett.10.531

[5] Kobayashsi M. and Maskawa T. 'CP-violation in the renormalizable theory of weak interaction', *Prog. Theor. Phys.*, 1973; 49 652–657. DOI:10.1143/PTP.49.652

[6] Feliciello A. and Nagae T. 'Experimental review of hypernuclear physics: recent achievements and future perspectives'. *Rep. Prog. Phys.*, 2015; 78 096301. DOI:10.1088/0034-4885/78/9/096301

[7] Chang K. *CERN Physicists Find a Particle with a Double Dose of Charm.* The New York Times; 2017 [https://www.nytimes.com/2017/07/06/science/cern-quarks-charm-baryon.html]

[8] Miransky V. A. *Dynamical Symmetry Breaking in Quantum Field Theories.* World Science; 1993; p 15

[9] Higgs P. W. 'Broken symmetries and the masses of gauge bosons'. *Phys. Rev. Lett.*, 1964; 13 508–509. DOI:10.1103/PhysRevLett.13.508

[10] Wineberg S. 'Conceptual foundations of the unified theory of weak and electromagnetic interactions'. *Rev. Mod. Phys.*, 1980; 52 515–524 DOI:10.1103/RevModPhys.52.515/ Salam A. 'Gauge unification of fundamental forces'. *Rev. Mod. Phys.*, 1980; 52 525–538, DOI:10.1103/RevModPhys.52.525

[11] ATLAS collaboration. 'Observation of a new particle in the search for the Standard Model Higgs boson with the ATLAS detector at the LHC'. *Phys. Lett. B*, 2012; 716 1–29. DOI:10.1016/j.physletb.2012.08.020

[12] CMS collaboration. 'Observation of a new boson at a mass of 125 GeV with the CMS experiment at the LHC'. *Phys. Lett. B*, 2012; 716 30–61. DOI:10.1016/j.physletb.2012.08.021

[13] Matsumoto R., Abe K., Hayato Y. *et al.* Super -K Collaboration. 'Search for proton decay via $p \rightarrow \mu^+ K^0$ in 0.37 megaton-years exposure of Super-Kamiokande'. *Phys. Rev. D*, 2022; 106 072003. DOI:10.1103/PhysRevD.106.072003

[14] Afanaciev K., Baldini A. M., Ban S. *et al.* 'A search for $\mu^+ \rightarrow e^+ \gamma$ with the first dataset of the MEG II experiment'. *Euro. Phys. J. C*, 2024; 84 216–225. DOI: 10.1140/epjc/s10052-024-12416-2

[15] Mannel T. 'Theory and phenomenology of CP violation'. *Nucl. Phys. B*, 2006; 167 170–174. DOI: 10.1016/j.nuclphysbps.2006.12.083

[16] Peccei R. D. and Quinn H. R. 'CP conservation in the presence of pseudo particles'. *Phys. Rev. Lett.*, 1977; 38 1440–1443. DOI:10.1103/PhysRevLett.38.1440

[17] Green M. B., Schwarz J. H. and Witten E. *Superstring Theory*. Cambridge University Press; 1988

Chapter 10
Conclusion

This book introduces the development of modern physics based on the "expansion of observation area" and "reduction of measurement uncertainty." Physical laws are based on the phenomena that were previously observed. Once the observation area is expanded, new phenomena that cannot be described with previous laws can be discovered. Some discrepancies can also be discovered by the reduction of measurement uncertainties. The development of new physics is required so that new phenomena can also be explained. This book introduces the development of modern physics by observing "phenomena with far distant celestial bodies," "structure of atoms, molecules, and nuclei," "phenomena with ultra-low temperature," and "search of elemental particle or nuclei observing the phenomena with ultra-high energy."

In ancient times, they lived in a small area within 50 km, and they believed that the Earth was a flat. After the movement between Greece and the Middle East became active, the Earth was recognized to be a sphere from the change in height of the Sun and constellations. Until the seventeenth century, the planets were believed to revolve around the Earth because there was no discrepancy between Ptolemaic theory and observation, whilst the accuracy of clocks was not high enough.

The first revolution in physics was given by Galilei. He observed Jupiter, the Sun, and the Moon using his handmade telescope. This was the first observation of phenomena outside the Earth. He also discovered the periodicity of the pendulum's swing, which gave a drastic improvement to the accuracy of the clock. From detailed observation of the motion of planets, Kepler's law was established, from which Newtonian mechanics was established. The speed of light was clarified to be finite by observing it outside the Earth (propagation over long distances) using an accurate clock. The measurement of the speed of light was the basis for establishing electromagnetism.

At the beginning of the twentieth century, there was an expansion of the observing area. Atoms were clarified to have a structure with nuclear and electron clouds. The property of electrons was not explained by the previous Newtonian mechanics and electromagnetism. In the same era, the dual nature of light (electromagnetic wave and particle) was clarified. The idea of dual nature was expanded to all particles, and quantum mechanics was established. The theory of relativity was established based on the constant speed of light.

At the beginning of the twentieth century, there was also the expansion of the astronomical observation area. Until the nineteenth century, only a small fraction of stars in the Milky Way were observed, and the universe was believed to be immutable. After observing stars outside the Milky Way, the expansion of the universe was clarified, and the Big Bang model about the birth of the universe was established. Since the beginning of the twentieth century, astronomical observation has been performed not only in the optical region but also with infrared, radio-wave, ultraviolet, and X-ray regions. The information about quasars at far distant places has increased drastically.

After World War II, the atomic clock was invented, and the measurement uncertainty of the clock was reduced drastically. Relativistic effects on the progress of time were confirmed using atomic clocks.

The invention of the laser also gave a revolution to experimental physics. The slight difference in length can be measured using laser interferometry (for example, gravitational waves). The energy structure of atoms and molecules was studied in detail by laser spectroscopy.

The development of laser cooling gave us an opportunity to observe the phenomena with kinetic energy much lower than 1 mK. With ultra-low kinetic energy, the wave nature of atoms and molecules becomes significant, and atomic interferometry and Bose-Einstein condensates are observed.

Until 1969, the proton was believed to be an elemental particle that cannot be divided further. This is because the energy gap between different states of quarks bound in a small area is large; therefore, the transition to different states was not observed until the collision with ultra-high energy was observed. Protons and neutrons were proven to be the binding of three quarks. Quarks bound in a small area (proton radius) have kinetic energy much larger than the rest energy (derived from the uncertainty principle between position and momentum). Therefore, pair production of quarks and antiquarks is easily caused in protons and neutrons, and it is fundamental to causing the nuclear force. Particles with large mass (including Higgs particles) have been discovered in collision with ultrahigh energy.

There are still many mysteries that have never been solved. For example, the unequal abundance of particles and antiparticles has not been derived from the present theory. The identities of dark energy and dark matter (γ-ray from the pair annihilation of dark matter might have been discovered) are still mysteries for present researchers.

How can we expand the observation area? Now we have observed quasars at a 13.1 billion light-years. We might be able to observe some quasars at a further distance. But we cannot observe the universe before the era of electromagnetic transparency (380 thousand years after the Big Bang) with electromagnetic waves. But we might be able to detect gravitational waves from the Big Bang in the future.

It might be proof of the existence of extra dimensions if we can find a violation of the Newtonian gravitational law at a distance below 1 mm. However, measurement of the gravitational force between atoms is difficult because it is much smaller than the electromagnetic force. The search for the gravitational force between neutrons might be useful.

The reduction of measurement uncertainties of different transition frequencies might be realistic to be attained in the near future. If the transition frequencies of nuclear or highly charged ions are measured with the uncertainties of 10^{-18}, the variation in fundamental constants might be detected, and it might give important information about dark matter.

Index

www.ingramcontent.com/pod-product-compliance
Ingram Content Group UK Ltd.
Pitfield, Milton Keynes, MK11 3LW, UK
UKHW020704041026
15590UKWH00001B/466